MILITARY THINKING OF ANCIENT INDIA

MILITARY THINKING OF ANCIENT INDIA

Shekhar Adhikari

PENTAGON PRESS

MILITARY THINKING OF ANCIENT INDIA

Shekhar Adhikari

ISBN 978-81-8274-910-8

First Published in 2016

Published by
PENTAGON PRESS
206, Peacock Lane, Shahpur Jat,
New Delhi-110049
Phones: 011-64706243, 26491568
Telefax: 011-26490600
email: rajan@pentagonpress.in
website: www.pentagonpress.in

Printed at Aegean Offset Printers, Greater Noida, U.P.

CONTENTS

PREFACE

Generally, it is assumed by the Western world that India lacks strategic thinking. But it is not so. Looking back into the pages of history one finds several military writings of great political and military thinkers. They have shown the intimate relationship between political, social, economic and war strategies. It is true that the history of ancient Indian military thinking was the culmination of various social and political thoughts of great thinkers. But the problem is that no where their military thought is stated at one place.

In considering the moral, political and economic background of war, ancient thinkers in India too have focused on the subject of military affairs. Military thinking does not merely means views on military power but it also take into consideration the nature of kingship, state, society and foreign policy. Military power comes into picture when the state or society are threatened by internal and external threats. It, then, become the primary duty of political institution to protect it by use of force. Army or use of force was important because it was regarded as means to obtain social and political goals. The ancient art of war and politics, thus, become important not only from factual point of view but one can also deduce certain principles of war which are of universal nature and can be applied even today. As such, one could infer that social, political and economic orders of various kingdoms (state) had produced great thinkers in ancient India who had visualized the significance of the art of war in the growth of kingdom.

One cannot deny the importance of war and use of force as an important aspect of statecraft. One can cite the example of Kautilya who was the cause of Mauryan victory by destroying the Nanda dynasty. This Kautilyan philosophy had deep impact on later military thinkings. It meant that victory could be possible only through the successful conduct of war. It is also important to mention that till 9th or 10th century all the ancient political and military thinking followed a single line of thinking, based on Kautilyan theory of statecraft and war.

It is strange, but quite true, similar military development took place in the ancient Greek and Roman empire. Even Plato, Aristotle, Rousseau, Machiavelli, Karl Marx and others regarded force or violence as a means to achieve the desired goals. A world without war was even theoretically inconceivable so long as one was within the established order. This aspect has been extensively dealt in various chapters in the book. There is also no denying that ancient military writings were written in a systematic and scientific manner. But one thing became clear that decisive battles were dependent upon favourable conditions, time, better strategy, tactics, military organization, movement of the troops, foreign policy and diplomatic tools. The propagation of mandala theory in ancient times is still applicable in modern world. It means the six diplomatic tools applied by the kings were similar to that of Karl von Clausewitz's contention that war was the continuation of politics by other means. Similarly, on war, Karl Marx stressed on diplomatic, economic, psychological aspects and only as a last resort to war.

In the light of the above analysis one can readily agree in ancient India, statecraft and war was text on military culture. This strategic or military culture had to be protected by bala (army) and punishment (danda). This was necessary to instil fear among the subjects so that society could be saved from evils. This resemblance Machiavelli's statement when he says "where there are good laws there are good arms." Having both, society, as such, was protected from external and internal threats.

The purpose of writing the book on various thinkers in ancient India was to ensure the critics that India had produced military thinkers who could visualize the significance of war or use of force as means to achieve political, social and economic goals.

ACKNOWLEDGEMENTS

I wish to extend my sincerest and heartfelt gratitude to hon'ble Vice-Chancellor, Professor Rattan Lal Hangloo, for his constant encouragement and support. My appreciation is also due to Dr. Shalini Chawla, Prof. Diwakar Kaushik, Mr. Shankar Ghosh, Dr. Dhirendra Dwivedi and a host of research scholars for their expert inputs. A special thanks to Mr. Rajan Arya, Pentagon Press, for publishing the book. Last, but not the least, I am grateful to my wife, Mrs. Pratibha Adhikari, for showing great patience while the book got completed.

1

INTRODUCTION

Man by his character and culture is a political and social living being. Since ancient times, his model of corporate life has been in accordance with form of authority or control. Being a rational human being, man had always thought about how he can improve his corporate life and cultural standard? This is the reason that competition and cooperation, war and peace establishments have become an integral part of man's social, political, economic and cultural life.

Although war is a violent, cruel and condemnable act yet it has remained an unbreakable companion of human culture. War is not only a medium to maintain a state's security, consolidation of peace, but it also expands the geographical limits to the utmost to execute policies and to obtain various interests to use it in order to enhance political power, pride and prestige. In other words, the history of warfare is closely linked not only with political history but also with cultural and socio-economic history.

Moreover, the development of military power is an indication of political, social, economic and cultural growth of a nation. The important question is what about the growth of culture in an uncivilized state which is marked by anarchy, chaos, confusion and mass killing? In this case, such states would produce war-like race, based on war-like spirit and where the people will be deprived of military leadership. Military leadership requires a degree of development of the intellectual faculty that could be lacking in an uncivilized state. Thus, intelligence is an important factor in determining superior military character of a civilized state. Karl von Clausewitz calls it "genius". Hence, a study of the cultural pattern of a civilized social and political order within a set of geographical areas and conditions would no doubt give rise to the development of economic organization, political machinery and socio-economic system, religious code of conduct, language, education, fine arts and artistic achievements. Any threat to the cultural pattern of social and political orders would be met by use of force (war).

It is strange that western strategic experts claim India lacks military thinking. But this is not so. In ancient India, the importance of war and use of force was seen as a significant aspect of statecraft. Ancient military writings were written in a systematic and scientific manner. Decisive battles depended upon favourable conditions, time, strategy, tactics, army organization, movement of the troops, foreign policy and diplomatic tools. In other word military thinking was closely related to Indian culture, and the protection of it was the primary duty of the political institution. But before we proceed further it is important to examine the recent debate on Indian strategic or military culture. Hence, the term strategic culture needs further analysis and then to comment whether India had or lacked military culture.

In recent years, studies on national security, based on the development of tangible and intangible elements of national power, and international relations highlight the importance of military or strategic culture. In the 20th century, the term culture was well understood by anthropologists, sociologists and psychologists. In the 1940s and 1950s efforts were made by the anthropologists to show the relationship between culture and state behaviour. But today, culture in any form has become an integral part of security studies. As a result, various scholars working on national and international security studies have realized that culture of a country, to a great extent, influences the grand strategy and state's behaviour.

Hence, there are a large number of combinations of actors and types of interactions or situations (governments, armed forces, national economic, social or religious groups and even individuals' charismatic personality), which has shaped the state's strategic behaviour in the formulation of national and international security. Thus, one can say that strategic culture of a country operates at different levels. The manifestation could be seen from the tribal or group levels to the organizational, national and cultural levels.

Before, we proceed further it is important to explain the term 'culture'. The important question is that where was the concept of culture derived from? One can agree broadly that it comes from history and geography. The concept is dynamic. It can evolve, adapt and even change dramatically by the circumstances of a situation. Edward B. Taylor has stated "culture or civilization, taken in its widest ethnographic sense, is that complex whole which includes knowledge, belief, art, moral, law, custom and any other capabilities and habits acquired by man as a member of society."[1] Contemporary anthropologists do not arrive at a particular definition of culture. Some definitions stress on the materialistic side of culture and other emphasise the idealist side of culture. Anthropologists do agree on certain characteristics of culture.

Characteristics of Culture

Topical culture consists of everything on a list of topics or categories such as social organization, religion or economy.

- Historical culture is a social heritage or tradition that is passed from one generation to another.
- Behavioural culture is a shared and learned human behaviour of a way of life.
- Normative culture is ideal, values or rules for living a way of thinking, feeling and believing.
- Functional culture is the way how human beings solve their problems of adapting to surrounding environment or living together.
- Mental culture is complex of ideas or learned habits that inhabits impulses and distinguish people from animal.
- Structural culture consist of patterned and inter-related ideas, symbols or behaviour.
- Symbolic culture is based on assigned meanings that are shared by a society.
- War or military or strategic culture is the ability and the capability

of the people and the society to develop power or force and to have the political and social will for a full and effective use of that "power".

Strategic or Military Culture

It is in the context, the study of strategic culture becomes important in assessing the nation's capabilities and implementation of its policy in national security and international relations. The pertinent question is what the term "strategic" means? According to the Oxford dictionary, the term strategic means to achieve a particular purpose or to gain an advantage. In military term, it means getting an advantage in a war or war-related situation. Basically, the term strategic culture is a western notion. But in Communist states or even in India in early times the term military, not strategic, was more in vogue.

It is true that strategic or military culture could influence national security and policy and it was first seen in classic works of Thucydides, Sun Tzu, Kautilya and Clausewitz. It was these thinkers who propagated the idea of war and military strategy as a "test of moral and physical forces". The object of strategy was not merely to defeat the enemy in the battlefield but also to bring submission to its own will, i.e. elimination of the enemy's morale. Moreover, great military leaders have laid emphasis on both political and military leadership in a mobilized society.

Strategic culture, as pointed by Jeannie Johnson, consists mainly of a consensus on including culture as a variable in analyzing foreign policy and security decisions.[2] Johnson defines "strategic culture as that set of shared beliefs, assumptions and modes of behaviour, derived from common experiences and accepted narratives (both oral and written) that shaped collective identity and relationships to other groups and which determine appropriate ends and means for achieving security objectives."[3] Jack Snyder, to whom the term strategic culture is given credit, has defined it "as sum total of ideals, conditional, emotional responses and pattern of habitual behaviour that members of the national strategic community have acquired through instruction or imitation and share with each other with regard to national strategy."[4]

In defining the term strategic culture, variables like identity, values and norms become important. Every state has its own security dimension.

Identity tells us the state's intention, capabilities and aspiration in assessing her position in international politics. Values state those principles and materials goods that are negotiable and that are not. Lastly, norms help in understanding those means which are to be adopted than others in securing state's objectives.

Military culture represents the ethos and professional attributes, both in term of experience and intellectual set up that contribute to a common understanding of the nature of the war within military organization. As Michael Howard had suggested no other profession is as demanding in physical and mental terms as the profession of arms. The military culture demands that the military institution must not only take resort to use of force for the purpose of war but rather must estimate the impact of technology, tactical operation, societal changes and the role of leadership as they apply to war. Thus, military organization can never completely evaluate peacetime innovations and preparation until the audit of war itself in which fear, chaos, ambiguity and uncertainty dominate. Military culture, thus, represents the intellectual and spiritual capabilities of the army, navy and air force to come to grip with the business of preparing and executing war.[5] This is how military culture is deeply related to changes in societal culture, the advance of technology and the impact of leadership. In 1932, two of the Reichswehr's generals, Werner von Fritsch and Ludwig Beck, rewrote the German army's basic doctrinal manual "Troop leadership" which served as the basis for the combined arms battle doctrine with which the Germans fought the Second World War. The opening paragraph of that manual encompassed the fundamental cultural assumption of the German army:

- The conduct of war is an art, depending upon force, creative activity, scientifically grounded. It makes the highest demands on individual.
- The conduct of war is based on continuous development. New means of warfare can call forth ever changing employment.
- Situations in war are of unlimited variety. They change often suddenly and are rarely discernible at an early point. Incalculable elements often influence the conduct of war. The independent will of the enemy is pitted against the Germans. Frictions and mistakes are everyday occurrences.[6]

In other words, the conduct of war or use of force has deep bearing on country's military ethos and institutions. Assessing Indian military thinking in ancient time one cannot easily co-relate with traditions, norms and values of its society. George Tanham in his famous work "India Strategic Thought An Interpretative Essay"[7] has too observed that how historical and cultural factors have shaped Indian strategic (military) thinking. It has also led to a debate within the civilian and military leadership and also tried to establish the relationship between technical, economic and institutional factors that have affected Indian military thinking. Further, he elaborates by stating that India is an extraordinary complex and diverse society and Indian policy makers have shown little evidence of having military thinking coherently and systematically about its strategy.

This led George Tanham to believe that Indian strategic culture had predominantly defensive strategic orientation. The reason was that the Indian people knew their own local history until 19th century. This observation was based on the idea that India throughout the course of history lacked political unity and hence did not thought in terms of national defence planning. Furthermore, he gives references where foreign incursion into India through the north west passes were seldom contained or defeated. As such, India did not had forward strategy and Indian forces were compelled to wage defensive battles. Hence, he concludes that lack of an expansionist military tradition was based on history of non-aggression and non-expansion outside the subcontinent. Thus, he concludes that India lacks strategic thinking or culture.

One should be well aware of the fact that ancient Indian military thinking was the culmination of various social and political thinking or ideas emerging within Indian society. The political structure was such that it was not conceived of one complete state or country but India in ancient times it comprised of various small, medium and powerful kingdoms. This has led to many critics to argue about the validity of the military thinking pertaining towards the development of art of war.

As to the moral, political and economic background of war, Indian ancient thinkers too have focused on the subject of military affairs. It is true that military thinking is closely related to military power of a state but at the same time there were sufficient evidences to prove that Indian military power influences the pattern of society, state, king and even foreign policy, and diplomacy was seen as an instrument of war. The use of force, kingdom, economy, and military power then, become important because it was the primary duty of political institutions (kings) to protect the society and the kingdom from various internal and external threats. Hence, one sees an intimate relationship between society and military power. Bala, then, became a powerful instrument in obtaining social and political goals.

Traits of India's Military Culture

According to Rodney W. Jones Indian strategic culture can be labeled as an omniscient patrician type.[8] This phrase means in Sanskrit *bharat jagat guru, or "India: The World Teacher."* Other forms of strategic culture could be theocratic, the mercantilist, frontier expansionist, imperial bureaucratic revolutionary technocratic and marauding or predatory. A description of each element of the philosophical and mythological factors that form the foundation of this culture is stated by the following chart:

Table 1. Traits of India's Omniscient Patrician Strategic Culture[9]

A. **Philosophical and methodological foundation**
   - Sacredness permeates Indian identity
   - Goals are timeless, not time bound
   - India's status is a given, not earned
   - Knowledge of truth is the key to action and power
   - World order is hierarchical, not egalitarian

B. **Instrumental implications**
   - India's external visage is enigmatic
   - Self-interest expressed externally is impersonal and absolute
   - Contradictions in the real world are natural and affirmed
   - Force has its place, but guile may trump force
   - Actions have consequences, good intent does not absolve injury
   - Entitlement inhibits ordinary compromise (hard to split differences, truth is not at ease with quid pro quo)
   - Compromise easily viewed as internal defeat (ephemeral, bends truth, dents sovereignty)
   - Trust is in right knowledge and action, is impersonal, and hard to build or replenish
   - Security is sedentary (encompasses a geographic setting and way of life)
   - Strategy is assimilative (appearance changes, reality is constant)

From the above analysis Indian military thinking has been drawn from various threads of its ancient civilization and larger political culture.

Scholars tend to equate war and peace in ancient Indian thinking to the real politik side of the religious spectrum. The great epics—*The Ramayana and Mahabharata*—deal with struggles between good kings and evil forces. In such a struggle good prevails over the evil forces. These epics are drawn in the context of kings, courts and opponents that lead to wars. Resort to force is treated as natural. In some cases good kings using extraordinary weapons like thunderbolt could well be connected to today weapons of mass destruction. Thus, the war was seen as a deadly combat between different forces and was even accepted when good triumphed over evil forces.

The most important ancient treatise on statecraft was Kautilya's Arthashastra. It is embedded in the same composite culture and reflects a period of the Indian subcontinent, which was largely dominated by Hindu kingdoms. Political situation in most of the kingdoms was responsible for chaotic state of affairs. To bring peace and stability, the necessity of military force was required. For example, the Mauryan overthrew mighty Nandas from powers and compelled other powerful kings to accept the suzerainty of Chandragupta who had acquired Punjab from Alexander. The boundary under him stretched from Gandhara to Bihar.

The ancient Indian writers had great resemblance with Greek thinkers like Plato and Aristotle and with Machiavelli of modern time in propagating the theory of use of force in critical conditions. Plato visualized that the legislators should focus on two principles – the people and territory. Aristotle went one step ahead by stating that legislation should take into account the neighbouring states and for this purpose a state should possess military force so it can defend itself from internal and external turmoils. Furthermore, Aristotle stressed on the states need to have wealth, freedom, justice and valour. Similarly, the Italian thinker Machiavelli too laid emphasis on Prince's central authority which should be backed by strong national militia. Thus, military power became essential for the survivability of the state.

As in western world, in ancient India too the primary concern of the kings were the kingdom and people's interests. There was a need to produce maximum military power. Ancient Indian thinkers observed that ally looked to the security of his own interests in the event of internal calamities and in the growth of enemy's powers. The decision to go for war or remain at peace rested upon self-interest or advantage of one's kingdom. Like the Roman and Greek empires the growth of Indian empire rested on war, conquest and profits.

The western notion of pacificism in Indian art of war could be well judged by the nature of fighting. Generally war could be offensive and defensive. Offensive action is one when army is in a superior and comfortable position. Defensive action means a weak king who is holding a inferior position. In modern age, Clausewitz has rightly said that both offence and defence are stronger forms of war. If powerful then an army should launch an offensive attack while weak should preserve its army and the subjects. In Harshacharita, it is mentioned that in ancient India King Rajyavardhana waged battles not merely for offensive purpose, but it was a combination of both offence and defence. Offensive action of the king could be seen in the form of "Chakravarti" or "digvijya" king or "would be conqueror". According to Rapson, it was always regarded as a legitimate object of the ambitious king of every dynasty to attain the position of Chakravartin or supreme monarch. In the western world, a similar notion was adopted. Here it is important to quote Adolf Hitler who wrote in his *Mein Kampf* that the true sphere of German colonization was not North Africa but the continent of Europe.

This demonstrates that ancient India followed a single line of military policy towards the adversaries. It included every aspect of war-violence, non-violence, strategy, tactics, organization of army, weapon system, marching, military leadership and so on. These have been extensively covered in the book.

The concept of military thinking in ancient India should not be considered a comprehensive explanation of a state's strategy. Though it is true the concept is related to security and it can define policy boundaries and assumptions but cannot always determine concrete policy choices. Perhaps, the reason for this is the changing dimension. Security is a dynamic concept. It changes due to the growth of industrialization and development of science and technology. The security issues have become more complex than before and all these issues are well related to cultural aspects of nations. Thus, the growing interest in ancient Indian security affairs remains the core factor in the study of state security and inter-state relations. There is one area even today where strategic analysts have understood the importance of the concept of peace when nations are talking of deterrence, strategic nuclear doctrines and diplomacy.

REFERENCES

1. E.B. Taylor, Primitive Culture, Research into the Development of Mythology, Philosophy, Religion, Art and Custom, 1958, Vol. 1, p. 1.
2. J. Johnson, Strategic Culture: The Theoretical Construct, Comparative Strategic Cultures Curriculum, SAIC, 31st October, 2006, p. 3.
3. Ibid.
4. See A. Johnson, Thinking About Strategic Culture, International Security, 1995, 19 (4), p. 36.
5. W. Murray, "Does Military Culture Matter," *Orbis*, Winter, 1992, p. 135.
6. Troop leadership, trans of Die Truppen Fuhrung, Washington, War Department, 1936. p. 1.
7. See G.K. Tanham, India Strategic Thought: An Interpretative Essay, RAND, National Defence Research Institute, Santa Monica, USA, 1992, pp. 1-82.
8. R.W. Jones, India's Strategic Culture, Defence Threat Reduction Agency Advanced Systems and Concepts Office, Comparative Strategic Culture Curriculum, United States Government Agency, 31 October, 2006, p. 4.
9. Ibid.

2

MANU

Introduction

The history of ancient Indian strategic thinking is the culmination of various social and political thoughts or ideas of great thinkers. Nowhere their strategic thought has been stated at one place. Hence, today, critics argue the validity of their thought pertaining to the art of warfare in ancient India. In considering the moral, political and economic background of war, ancient thinkers in India too have focused on the subject of military affairs. Strategic thinking does not merely mean views on military power but it also deals with the nature of society, state, kingship and inter-state relationship. Military power comes into picture only when the society and state are threatened and, then, it becomes the duty of a political institution (kingship) to protect it from internal and external aggressions. In other words, there is an intimate relationship between society, state and military power. As such, one could infer that social, political and economic orders of kingdom (state) had produced great thinkers in ancient India, who had visualised the significance of the art of war in the growth of society and state.

Manu's works on the art of war have proved immeasurably fruitful in shaping the history of ancient India. His work Manusmriti is a work of pre-eminence not only to the students of ancient India but also to the students of modern world. His work is also known as Manu's Dharmasastra which is a "work of encyclopedic scope."[1] About the exact date of its composition, it is generally believed that the book was composed during 200 B.C. to 200 AD. The name, Manu, can be traced from the Vedas. A Vedic poet prays that he may not be led away from the ancestral path of Manu. Another Vedic passage (Rig Veda X, 63-7) says that Manu was the first to offer sacrifice. In Taittiriya Samhita (II, 2.10.6) and Tandya-Maha-Brahamna (23.16.17) state "whatever Manu said is medicine". Taittiriya Samhita (1.5.6) also says that mankind is Manu's. This means Manu is the creator of the Hindu social order. Vasista Dharma Sutra refers to Manu by name in various places or even quotes his views. Even in Mahabharata and Kautilya's Arthsastra there are references of Manu in many places.[2]

Manu has dealt in detail on the social, political and judicial conditions of that period. It includes social obligations and duties of various castes of individual in different stages of life. As the most authorative work on Hindu Law, it presents the normal conditions of Hindu society and civilization. As a traditional lawgiver of antiquity, Manu has been compared to Hammurabi, Moses, Lycurgus and Solon. But the real credit goes to Manu who actually realises the importance of socio-political institutions of the land and civil and criminal law.

If this was the general social, political and religious conditions of that time the pertinent question was how secured was the people, society and the kingdom (state). These issues have been taken up by Manu emphasising the role of kingship in protecting its people and the state from internal disturbances and external aggressions. To have a deeper insight of Manu's military thinking one has to reassess the nature of society, state, political institution and jurisdiction (danda).

Society

Manu lays much emphasis on the welfare of the society. As such, security of the society was related to the welfare of the people. For this he emphasises two terms *varna* and *ashrama*. The objective was to secure the hereditary qualities and also utilize the prevailing environmental conditions. Varna divided the society into classes: intellectual and manuals. The intellectual class was again divided into three classes and these classes were meant for developing the environmental conditions. Their works were meant for the highest efficient working of that particular group during the Ashram period or the natural period of human life. Sudras were allowed to enjoy their own types of native culture. As such, Hindu society was made up off our separate types of cultures, which had different functions to perform.

So far the culture of Brahman was concerned, it was based on high morality and was expected to speak the truth. The culture of kshatriya was to show courage during fighting. Though this class was also highly moral, he could tell a lie in war which was permitted by customs. The vaisya's life was more focused on commercial and trade transactions. A Sudra who does not possess wealth was accustomed to lead his normal life in accordance to its own customary laws. All these four types of varnas were important for the proper functioning of society. In other words, every person was allowed to enjoy his social work and status to the utmost and Manu gave them an incentive to work to make the society a perfect one. Later on, Manu in Chapter 10 of the Manusmriti divides this general population by giving them universal employment. Hence, for the first time, the class of avarnas also emerged and rules were made for them so that they can be accommodated in the society. In this way, Manu lays stress on five classes. It is also mentioned in the Atharvas Veda III, 24-3 which clearly states that these classes descended from Manu. Thus, the division of society into five classes or jatis of brahman, ksatriya, vaisya, sudra and avarnas was first pronounced by Manu. Manu's description of the social structure is also seen in Ramayana, Mahabharata, Puranas and Dharmasastra. This type of social structure continued till the advent of Buddha.

It appears that the division of society into five classes has also led to division of works. This could have produce social inequality in the society as it is found in every modern civilized society. But in actuality, efforts were made to remove this inequality and make them equal. But this seems to be an ideal principle because we can fight for equality and liberty but history has proved that we have failed in many occasions to obtain the right of equality and liberty.

Manu's emphasis on the nature of society has also led to analyse the importance of panchayat during ancient time. Today, we call it village panchayat. The term denotes that it was an elected institution in order to carry out its functions democratically. About its origin, it is quite debatable.

As Kelkar opines that the word 'Madhyamashi' in Vedic literature signifies the institution of panchayat and it must have existed during Vedic times.[3] The term panchayat is made of two words *panca* and *ayatanam*. The word panca means five and ayatanam denotes an abode. Hence, it means an abode of five persons. Panchayat may be taken to mean as an abode of the brahmanas, kshatriyas, vaisya, sudras and avarnas and all these were the creation of Manu.

Every village elected its own panchayat. It comprised of elected representative from every groups. It was the governing unit of the village and formulated its own laws, rules and regulations. In case of any conflict between the Rudhi law of the family, jatis or village with that of the Smriti Law or the Central Law, then the Rudhi Law or the Law of the people were given preference over the latter. Such laws enabled all the castes to perform the function in a democratic way by enjoying the rights of equality and liberty. Thus, full enjoyment of the democratic sovereignty of the people and liberty without sacrificing equality in any way was, perhaps, one of the major achievements of the Manu's Hindu society.

If liberty, equality and sovereignty of the people are the true measures for ideal democracy, the social structure of Manu satisfies, then, in all the spheres. It is, therefore an "ideal democracy". And as such its head is elected and approved by the people, it is also Republic. Therefore, the social structure of Manu can be safely described as the "Ideal Democratic Republic" of Manu.[4]

State

Next to society was the importance of state as visualized by Manu. There was intimate relationship between society, kingdom and kingship. Here, we will see that the term rastra is used which denotes a kingdom because the concept of state and rastra (nation) are modern in nature. For our purpose I will stick to the term kingdom rather than rastra or state. In Manusmriti there are references of the kingdom and it is dependent on the king.[5] The king should use four expedients. As the weeder plucks the weeds and preserves the corns, similarly the king defends the kingdom and destroys his opponents. An oppressive king will be deprived of his life and kingdom. Only by protecting his territories, the king increased his happiness.[6] Thus, J. Basu comes to the conclusion the conception of the state was closely related to the evolution of the kingship.[7]

The important question is in what manner the kingdom was created? There is no definite answer to it because it is difficult to trace the exact beginning of ancient Indian society. But there are references in Hindu literature that the kingdom in ancient India emerged within the society. There was some kind of interaction between king, religion and economic needs of the subjects. These needs were responsible for the formation of kingdom. In addition to these factors, B.S. Saletore believes that there must have been the tendency to subordinate an individual to an authority which looked to the interests of the larger group and which tried to unite all the groups within it.[8]

An interesting theme of our discussion is the relationship between religion and kingdom. Do the relationship was the same, as we saw in Medieval European history where Church was considered "state within state". In ancient India, like Egypt, Greece and Rome, religion existed long before the kingdom. It was an age of an undifferentiated religious-political society where the authority of the priest was greater than that of king. This was probably because the priest was the undisputed interpreter of law and the king relatively occupied a subordinate position in society. The perception, today, holds no good because religion is a part and parcel of the functioning of the state. The rise of modern nation-state began with the disintegration of the medieval structure based on religion.

In those days of the Vedas, the Samhitas and the Brahmanas, ancient Indian society was well aware of important concepts relating to their socio-political life. These concepts were *rajya, svarajya, samrjya, bhaurajya, vairaja, maharaja and adhipatya.*[9] Whatever may be the term we use in case of kingdom one thing becomes clear that brahmanas were excluded from exercising the king's power. But it was found that "the king (rajan) doubtless belongs the rajasuya; for by offering the rajasuya he became the king: and unsuited for kingship was the brahmana. Moreover, the rajyasuya was the lower and the vajapeya was the higher sacrifices. For by offering the rajasuya he became a king and by the vajapeya he became the emperor (samraj). The offer of the king is lower than that of the emperor. Thus, the king who performed the vajapeya became an emperor and possesses himself of everything here."[10] This indicates that a ksatriya which performed rajasuya sacrifice became an emperor. As such a ksatriya was to attain the position of a king while the kingship was denied to brahmanas.

Function of the Kingdom

The first important function of the kingdom was to protect and see that all varnas followed their respective duties. Any defaulter belonging to the sudra or vaisya varnas would be compelled by the kingdom to perform their duties.[11] The people were allowed to follow the common laws. Even the king was not spared by Manu from not obeying the law. When a common man would be fined one karsapana, the King shall be punished by paying 1000 panas.[12] The kingdom's function also included the establishment of peace so that the subject are freed from external pressures.[13] It enacted laws for the control of prices of the commodities. The prices of the commodities were fixed by the kingdom so that the businessman did not charge exorbitant prices.[14] It also made arrangements to settle disputes among groups, families and guilds pertaining to properties, boundaries, deposits, debts, contracts etc.[15] The kingdom also maintained a police to check fighting among groups. It promoted the cause of education by giving gifts to the educational institutions by honouring teachers and by providing opportunities for the educated young men.[16] The kingdom also punished those who used to indulge in various types of crimes.[17] Thus, the scope of the functions of the kingdom was very wide. Motwani, while analysing the functions of the state, stated, "some of the legislations would seem almost socialistic to a student of modern political science".[18] There is some element of truth with regard to Manu's thought because the legislation of Manu do have relevance in the context of modern laws and ordinance.

The Constituent Elements of the Kingdom

Manu mentioned seven limbs of the kingdom and these constitute seven parts of a kingdom. These were the king, amatya, janpada, bala, fort, kosh and mitra.[19] It is also known as the Saptang theory to which both Manu and Kautilya had subscribed. Emphasizing the significance of each part, Manu stated that all the elements were important and the destruction of even one element would be disastrous to the kingdom. There was no single part dominating the others because all these elements had their own inherent qualities. To Manu, each part was preoccupied to accomplish certain objects and that objects were affected by its means. Each element had a fixed duty. In other words, all the seven limbs that comprised that kingdom constituted its "living and growing organism". This theory was further given impetus by B.K. Sarkar, Rangaswami Aiyangar, Dr. Jayaswal and Dr. V. Kane. The relative importance of the seven of limbs of the kingdom has been discussed below.

The King or Swami

The king or swami occupies an important position because he constituted the most vital organ of the kingdom. Regarding the appointment or election of the king it is found that in Vedic literature "Kingship seems often to have been elective."[20] To V. Kane "the king was the state in ancient India".[21] The king was the source of social progress and prosperity and everything was depended upon the king. Manu looked upon the king as a protector of the people's and stability of the social order. It was well accepted that if the king was virtuous, the kingdom prospered. If the king was vicious and tyrannical prosperity of the society declined and the people suffered heavily. In Ishwar Tope's words, kingship goal was the goal of the state. The kingship was the moving spirit and the state was its expression. It embodied the basic ideas of the state. One without the other was inconceivable. The life of kingship was the kingdom. Both were inseparable. Coexistence was their law. The welfare of the state depended upon the kingship.[22]

The king has been rightly called as the highest and most important organ of the kingdom because he was the source of strength and unity. To him, unity of the kingdom had to be secured at all cost. The king was responsible for the efficient and smooth functioning of the government machinery. He was highly respected by the subjects and was seen as both "father and son".[23] Since the king was the living symbol of the unity of the kingdom, every attempt was made to strengthen his position.

Manu proclaimed that the Lord had created the king with essential parts taken from Indra, Varun, Agni, Sun, Moon and Kubera. The statement means that the king was endowed with the qualities of these gods. It also meant to magnify his status and importance amongst the subjects. The Agnipuran clearly states that the king exercised the functions of nine deities.

The above analysis shows that Manu envisages the concept of divine origin of kingship. Prof. V.P. Verma remarked that "Manu had in mind the creation of the human king by God out of the eternal particles of eight gods. He did not solve the logical difficulty involved in the notion of the particles. If the particles of gods were non-physical, a physical human being could not be created out of them. If the particles of god were physical they ceased to be gods and they became considerably human.[24] Enumerating his qualities, Manu says that like the sun, he burns the eyes and minds of those who gaze at him. Fire burns only the guilt but the King's wrath burns the entire family along with property and prosperity. He declared that the goddess of fortune dwelt in the king's favour, victory resided in his valour and death abided in his anger. The king will also destroy those who go against him.[25]

The security of society and the kingdom by the king from internal and external threats will be discussed ahead.

Amatya

Next to King was amatya or minister. In actual practice, the king had number of ministers to help and advise him in conducting the affairs of the kingdom. In addition to amatya, two other terms: sachiva and mantrin were also used. For all these, amatya is the oldest. There are different opinions regarding the constitution and size of the council of ministers. According to Manu, it should consist of 12 members. Later on, Kautilya felt that the exact strength was to be determined according to the requirement of the time. It was difficult to preserve the secrecy of the decision if all the members were consulted. Manu, then, recommended that in such cases consultation of three or four members was sufficient. In other words, he favoured the setting up of inner cabinet consisting of a few trusted and experienced ministers.

Janpada

The third limb was known as janpada. The term corresponds to the two element – land and population – as mentioned by the western writers. Like Plato and Rousseau neither small nor big states were favoured as an ideal. What was important was the quality of the soil.

In Manu's book we find an improvement in the concept of territory. Hence, unlike in west, janpada during Manu's time was marked by certain characteristics. Manu stresses "Let the King settle in a country which is open and has a dry climate, where grain is abundant, which is chiefly inhabited by Aryans, not subject to epidemic diseases (or similar trouble) and pleasant, where the vassals are obedient and his own people easily find their livelihood".[26] Also important for the King was that he should strive to gain what he had not gained and carefully preserve what he had gained.[27]

Durga

The next limb in the list was durga or fort. The reason for the construction of durga was the nature of topography and geographical situation in northern India. The fortified capital became important for the security and safety of the kingdom. The necessity of construction of durga was also related to time factor. Manusmriti speaks of royal fortresses. The king was to build in his kingdom "a town making for his safety a fortress, protected by desert, or a fortress built of stone and earth or one protected by water and trees or one formed by an encampment of armed men, or a hill fort. He mentioned of six different kinds of fortresses. The first three are said to be inhabited by wild beasts, animals living in holes and acquatic animals and the last three were inhabited by monkeys, men and god respectively.[28] In a later context, Manu believes that once the king has settled his kingdom and constructed forts in accordance with the institutes, he, then, should remove dangerous men who are like thorn to his side. Manu in defence of forts, said "a single archer under the shelter of the fort can fight hundreds of the enemy and a hundred can fight 10,000."

Kosa

Manu and later on Kautilya held the same view on kosa. They looked upon the king to guard his finances and keep the treasury intact. Even earlier, in Vedic times we saw merely the beginning of the theory of finance. The important question was what was the source of public finance? One of the sources of revenue was the collection of resources from war. The word udaja occurs in the Maitrayani Samhita, which is regarded by Prof. Delbruck to mean share of the booty taken by the king after winning the war. The other was bali, a term used in Rig Veda frequently. Bali means voluntarily offerings by the subjects in the form of gifts and tributes.

Both Manu and Kautilya made certain provisions to guide the king in his taxation policy. He could levy only those taxes that had been sanctioned by the smriti writers, i.e. by sacred cannon law and at rates fixed by them. Manu even permitted the king to take a sixth part of the grain crops or the produce of the soil. Later thinkers, however, permitted the king to take even 1/3 or 1/4 in time of emergency or national distress. Whenever the king had demanded taxes at higher rate he had to take the consent of the people.

Manusmriti further mentioned a detailed list of the principal sources of state revenues. The main source was taxation. Taxes were levied on the produce of land, cattle, trees, meat, honey, clarified butter, perfumes, medical herbs, substances used for flavouring food, flowers, roots, fruits, leaves, pot-herbs, grass objects made of cane or earthen vessels and all other articles made of stone. Tolls were another source of income. It was levied on common people carrying goods from one place to another. But no taxes were taken who were doing manual labour for their livelihood.

Bala

The sixth limb of the kingdom was bala or army. Some writers even called it danda. But, bala is more expressive that it denotes military power and danda includes punishment. Bala was generally associated with the protection of forts from any types of aggression. In Rig Veda, the Vedic kings defended their territories by fighting battles into the enemy's territories. Hence, bala became important in the defence of the kingdom. The duty of protecting the territory was given to the kshatriya class.

The earliest ideas of bala and warfare seem to have undergone a change during the time of Manu. Manusmriti not only mentions the duties of the kshatriyas in the matter of carrying arms for striking and of throwing missile but also a new beginning had started in the history of inter-state relations by laying emphasis on the aggressive designs of the foes and then pushing the war into the enemies' land. Manu states "of him who is always ready to strike the whole world is in awe...."[29] Not content with this, Manu advocates the theory of rajmandala and makes it incumbent upon the king to undertake an aggressive war. The theory of rajmandala as expounded by Manu will be taken later on.

Mitra

The seventh and last limb of the kingdom was mitra. The inclusion of mitra as an important part of the kingdom by the ancient Hindu writers throws much light on the concrete nature of their approach to the subject. For them, kingdom was a concrete entity which had close relations to other entities of the same nature. It showed that their approach was external and not completely internal. This ancient idea is also expressed by western thinkers in modern age when they refer to the various elements of the state.

According to Manu, "The king did not prosper so much by gaining gold and treasury as by obtaining a friend." He, further, added that mitra though a weak one was greatly praised who was righteous, grateful, whose subjects were well contended, who was loyal and who pursued the work upto its completion".[30] Thus, a mitra in the form of a neighbouror distant king was very important for a king. On this basis, Manu propounded the theory of rajmandala.

History had proved from time to time that mitra had helped a kingdom in defeating the enemy. For example, with the assistance of the German forces, under Blucher, Wellington was able to defeat Napoleon in the battle of Waterloo. The British forces in India with the assistance of Indian allies and friends defeated the Marathas and the Sikhs. In the 20th century, the American assistance enabled the Allied power to crush the ambitious design of Axis powers in World War II.

The Theory of Justice

The King performed its duty of protecting the society and the kingdom through the method of coercive enforcement. It was called justice which was based on the concept of 'dharma'. The term dharma does not only mean the scared laws but, in addition, it includes: (1) law, usage and custom; (2) moral merit or virtue; (3) duty; (4) right, justice; (5) piety; (6) morality and ethic; (7) nature or character; (8) an essential quality, characteristics property or an attribute.

Justice, in the Indian context, was a human expression of wider universal principles of nature and if men were entirely true to nature their action would be spontaneously just. Impartial administration of justice was always regarded as one of the main duties of the king and he was considered as the fountain of justice.

The ethical code of a society and its cultural standards were interrelated. In the early Vedic times justice was administrated by tribe and clan assemblies and the juridical procedure was not rigid. But with the extension of the functions of the kingdom and the growth of the royal power, the king came gradually to be regarded as the fountain of justice.[31] As such, a more or less elaborate system of judicial administration came into existence.

In Manusmriti there are twelve chapters devoted on justice. It could be divided into four. These are *acare, prayascitta, vyavahara and rajdharma.* In the book, references of legal procedures could be seen upto eight chapters. In the eighth chapter Manu's expertise in giving direction was related to the maintenance of law and order. Manu's king appeared as a clever advocate capable of managing the disputes of common man.

Besides justice, the term danda or punishment has been referred by Manu. Normally, the term means coercion. He believed that the king must know dharma and danda. Dharma is important for the proper maintenance of its rules and those violating these rules are punished severely. He makes it clear that one of the duties of the king is to protect the life and property of his people by applying the theory of danda. Manu assigns to him the same divine origin as to the office of kingship. The lord created danda before he appointed a king so that the king could discharge its duties properly and efficiently. Danda is considered as the protector of all creatures and law. According to him, danda rules all people and protects them. Through the fear of danda, criminal tendencies were prevented even when public was asleep. Thus, it was danda that kept all classes of the society or varnas and the ashramas within the limits of discipline. Manu also applied the doctrine of universal jurisdiction of the king's danda over his subjects. Thus, he laid down the principle of the king's unlimited jurisdiction over offenders irrespective of the rank, status or relationship. He felt that the king's mode of application of justice was key to the prosperity and destruction of the individual and the community.

It seems that well-established jury system prevailed during Manu's period. As per rule, the king was to preside over the court when justice was delivered to the people. The court comprised of three judges, including the Chief Justice. He even recommended that the king delegates the power of judicial administration to a brahmin in his absence because the appointment of brahmins and experienced councilors would help the society in accomplishing the tasks.

Manu had also provided the qualifications of the judicial administrator. The one who was truthful, acted after due consideration, wise and knew the respective value of virtue, pleasure and wealth could be the judicial administrator. A king who properly inflicted punishment prospered with respect to those three means of happiness but if he was cunning, partial, deceitful he was to be destroyed, even through the unjust punishment which he inflicted. Manu observed that the judicial administration should not rest in the hands of a feeble-minded king.[32] If judicial administration were given to such a king he would ruin the whole kingdom.

Manu states that there are eighteen types of legal suits. These are: (1) recovery of debts; (2) deposit and pledge; (3) sale without ownership; (4) concerns among partners; (5) resumption of gifts; (6) non-payment of wages; (7) non-performance of agreement; (8) rescission of sale and purchase; (9) disputes between owners of cattle and herdsmen; (9) disputes regarding boundaries; (11) assault; (12) defamation; (13) theft; (14) robbery and violence; (15) adultery; (16) duties of man and wife; (17) inheritance and partition; and (18) gambling and betting.

In the light of the above analysis of society and kingdom, it becomes essential to analyse his view on war and its impact on the state. What does Manu thinks of society and kingdom in terms of security and stability? Because every society and kingdom needs a certain amount of security so that the subjects may be free to enjoy their complete life without any threat to them. To discuss these issues what is evident from Manu's writing on the security of kingdom in terms of internal and external sources. It is security which makes a society flourish, perish, or even establish new empires. In assessing this pertinent issue, it becomes important to reassess and properly understand the nature of the king, his duties and functions (both non-military and military) so that the society and kingdom could be protected from internal disturbances and external aggression. Equally important for us is to discuss the political institution (kingship) in detail and then try to correlate his style of functioning with modern strategic writings.

The King

According to Manu there is only one yardstick to assess the development of a kingdom's powers. To Manu, the kingdom was ruled by the king who belonged to the ksatriya varna. The kinship was hereditary but it was responsible to the people. There was an absolute king of the ksatriya class who reigned due to the divine right.[33] Even as an infant the king must think that he was mortal, for he was a great divinity in human form.[34] According to Manu, the creation of a king by the Lord took place in order to ward off the fear generated by anarchy among the people.[35] The king was formed out of the eternal particles drawn from eight guardian deities, Indra, Vayu, Yama, Surya, Agni, Varuna, Chandra and Kubera.[36]

Functions of the King

The functions of the king could be categorised into two broad heads: (1) military and (2) non-military. Both are important for the king to provide security and stability to society and in that sense both are related to each other. If the king fails by non-military methods then forcible measures adopted by him will ensure the security and stability to his society and kingdom.

Non-Military Functions

Under non-military functions Manu lays much emphasis on executive, judicial, legislative, ecclesiastical, revenue and enlightened duties of the king.

Executive Functions

It is quite evident from Manu writings that the King had great executive responsibilities in meeting the internal and external challenges. Thus, to the king the protection of society becomes important because he had to protect his subject with great zeal and vigour. Unworthy person were punished. By doing this, his kingdom flourished like a well watered tree.[37] If the king failed to punish the unworthy then Manu says the "stronger would roast the weaker like fish on a spike".[38] Like in every society, the protection of the weak against strong assumes much importance. For this if the king had to resort to arms he should do so because the security of the society is his first priority.

Administrative Function

Under this, comes the proper execution of his policies and plans. To implement this, the king even appointed ministers and officers. The king also appointed intelligent supervisors who were to inspect all the act of these men who transacted the kingdom business. By doing so, he wanted administrative efficiency so that the people could enjoy happiness and hence prosper under the able king.

Revenue

Revenue is an important source of income. Without proper utilization of the revenue the kingdom could be put in shambles. It was the duty of the king to fix the rates of taxes. The king did so much after much considerations, so that the common man were not affected. In other words, lack of proper collection of the revenue the king would not be able to satisfy his subject by providing all the facilities and comforts to them. These taxes, as we see it today, is the backbone of the state economy.

Hence, by paying taxes every member of the society enjoyed the individual sovereignty as well as liberty and equality of democracy of civil, social, political and religious nature. As every person was given vocation according to his own inherent qualities with a guarantee that the basic needs of the subject will be satisfied and as such economic equality will be achieved. Hence, people enjoy full equality without losing his liberty.

Today, in a capitalistic society there is liberty but no equality. In contrast, the communist states emphasize on equality while denying the right of liberty. But, during Manu's time, Hindu society could achieve both to the best possible extent by extending their services without hindering the prosperity and security of the kingdom.

Inspite of the non-military functions of the king there was customary division of powers. All these branches pertained to administration. The reason was that laws or rules regulating the life of the subject and social culture were either laid down in the Smritis and other law books. As such the king had no power to alter the old and make new rules. But still, the king, according to customary law, had to perform those functions in such a way that the kingdom prospered.

Other Duties: Besides these important duties and functions of the king, the king, for his own welfare and his own subject, exercised ecclesiastical duties by performing sacrifices. The king need to offer various sacrifices at which liberal fees were distributed. In order to attain merit, he should give brahmins enjoyment and wealth.[39] Hence, the king honoured and supported the brahmins by offering gifts and maintenance allowance.

As discussed above, the king was the supreme judge. Trail of the cases were begun by the king with a collective mind.[40] The decision had to be just so that no one suffers. In his absence, the judge pronounced the judgment and if something went wrong with the judgment the king had the power to reverse the judgment. In this case, the king was, infact, the upholder of the cause of justice.

Thus, the king believed in good laws as it will protect the society and kingdom from the evil forces. Later on, Machiavelli, the famous Italian political and military thinkers, stated where there are good laws there are good arms and where there are good arms there are good laws. Then, only the society and kingdom were to be secured.

On War (Military Power)

It is interesting to note how Manu looked at war and its relationship with the king so that the king and his subjects could be protected from internal and external threats. The word 'protection' is used to bring stability in the kingdom. This conception is a very ancient one. Prof. Macdonnell and Keith have shown that in the Vedic literature it was connoted by the term 'natha' which appear 'neuter'. But in the post-Vedic period it meant in the masculine sense meaning protection and this was very common.[41]

"To fight to death for these great passions (defence of family's honours and country's welfare), to feel that it is honour to die for them and dishonor to retreat from them, to give up everything material for them, to war for them to the last man, is just, true and righteous war and there is not a thought or a passion connected with it which does nor exalt and ennobles, not only the persons and the nation which wage it but the whole body also by humanity to whom the great tradition of this defence is handed down and whom the memory of it inspires, teaches and ennobles."[42]

Generally in ancient time, people looked upon king as a father. Prof. Hopkins has suggested "having a king as a father, was not altogether meaningless, but it was supposed that the king was considered a father since in all early societies, it was the father who alone was entitled to protect his children and dependents." Manu also states that the king had been created to become the protector of the varnas and orders, who all according to their rank, discharged their several functions. Like others, Manu opines that the king should behave like father towards his subject. The kingdom must protect the weak against the strong and prevent social confusion.

At certain places, we come across the term "force". The ancient thinkers primarily looked upon the concept as a primary cause of the origin of state. Force was important as an instrument to protect the kingdom. The ancient thinking laid emphasis on the propensity of man to encroach on his weaker neighbour and to be prone to commit acts of disorder and aggression. Prof. Salmond in his book "Jurisprudence" wrote man by nature is a fighting animal and force is the ultima ration not of all kings alone but of mankind." Similarly, Machiavelli opines that those who have discussed the problems of civic life demonstrate the use of force and in history there are ample examples to prove that who organises a state and arranges laws for the government must presuppose that all men are wicked and that they fail to show their national depravity whenever they have a clear opportunity.

The term force was also seen by Manu in a different context. As stated above, society had existed before the emergence of state. For the reservation of society force was used in accordance to dharmasastra. Today, force is the visible expression of the state to protect its various components. How a state could be protected? It could be protected by the king, army, the police and the legislative authority. Even in Manu's military thought these various methods are mentioned. To him, the king protected the kingdom not for his personal gain but for social welfare.

An important question that arises whether the word war, as understood in modern age, was used in ancient time. More important than war, the term battle is frequently used in ancient times. Today, when we talk of war, it is an armed struggle between two or more states with unlimited space and use of complete resources to win over his adversaries. Generally, in war, there are many battles but battle does not consist of various wars. This is important in the sense that in ancient India the writers and historians have used the word war. But for our discussion here war should be assumed in term of battle.

Manu, while taking note of military power of the supreme authority of the kingdom, believed that the protector should act justly in his own kingdom and should take harsh measures on the enemies. War, as understood by Manu, is said to be of two kinds: first when made of one's own accord, in season or out of season on account of one object and second, when made on occasion of injury to a friend.[43]

To Manu, expeditions are said to be of two kinds: first when the king marches alone when some affairs admitting of no delay have unexpectedly arisen and second when he is accompanied by an ally.[44] The unexpected expedition in modern times means 'sudden attack' emerging out of the fact to surprise the enemy or to demolish the enemy without the knowledge of the attacker.

Manu says the king should indulge in battle when a king is challenged by kings of equal, greater or less power. In this way, he can give protection to his subjects. While performing his duty, Manu says, that the king should not forget that he is a ksatriya. The chief cause of bliss or happiness of the king is that he should never cease battle, work for the protection of the people and should obey the brahmins. Manu opines that kings who are desirous to slay one another and fight with their greatest strength in battles without turning away go to heaven.[45] But Manu focused more on righteous war which was based on ethical or moral principles as regards to unethical war. War without moral principles degraded the institution into mere animal force. Hence, there came into being the moral code and code of lawgivers. A king desirous of dharmavijaya should confirm to the code of ethics enjoined upon warriors. Of the two kinds of war, the dharamyudha and the Kutayudhaya, Manu gave importance to the former because it was a just and righteous war which had the sanction of society.

Elements of the Army

It is also essential to make an assessment of the composition of army during Manu's time. A brief mention may be made of the progress in the history of the components of the army from Vedic times onwards. In those Vedic periods, the bow (dhanus) and the arrow (parusa) were the commonest forms of martial equipment. There is nothing in Vedic literature to denote that the Aryans had divided their army into fourfold division, although ancient historians emphasize on the existence of infantry (patti), chariots (ratha) and horses (asva). Although the elephant is mentioned under the name of ibha, gaya, naga, varana, sukladanta and hastin, yet its use in warfare is not apparent. But B.A. Saltore felt that the well known fourfold division of the army (infantry, horse, chariot and elephant) appears in the Manusmriti in which the different components are mentioned in various contexts.[46] The king, generally, inspects his fighting men, all his chariots and beasts of burden, the weapons and soldiers' outfits and garments.

Military March

A good army, under the king, will boost his confidence and even translates his ideas into actions. The king should appoint seven or eight carefully examined ministers who are hereditary, learned in the treatises, brave, skilled in the use of weapons and well descended.[47] With them, the king should be able to deliberate ordinary matters of peace and war, the condition of the kingdom, wealth, protection and the pacification of acquired territory. Thus, the army was depended on these ministers.

The king should march upon the road with the troops distributed like a staff or a waggon, or a boar, or a makara or needle or a garuda.[48] It means that "staff" is in straight columns. The "wagon" has a sharp van with a broad rear. The "boar" has a sharp van and rear with a broad centre. The "makara" (sunbeast) is the opposite of this, having a narrow centre with a broad van and rear; the needle is a long, thin, sharp-pointed row. And the garuda (mythological bird) has a very wide centre but is otherwise like the boar.

The army is spread out in a circle and the king is in the middle of it distributed like lotus. This has been done when he suspect danger and to meet outcome the king then, expand his forces. The king should also put his general and commanders in all directions and when he fears danger then let him arrange his forces to the front. If he has but a few soldiers they should be made into one compact body. If he has many, he should mass them or distribute them about as he chooses. The needle array has been explained above. The thunderbolt array is where the army is drawn up in three divisions by separating the two flanks.

The king should fight with chariots and horses on level ground, so with elephants on marshes; on ground covered by trees and shrubs with bows; and on clear ground with weapons, swords and shields.[49] The king having made arrangements at home and for the expedition in due form have secured a position, i.e. by winning over malcontents from the other side, he should secure a foothold in his enemy's kingdom. If the king has prepared three ways and sixfold forces he, then, gradually move by military rule to the enemy's city. The king should be more watchful in both the cases, firstly of an ally who secretly favours the enemy and secondly pretending to be a friend by going away but coming back by a secret way. Such allies or friends are enemies that are more dangerous.

Tactics

According to Manu, the king should develop his own of tactics by which he can successfully wage a battle against the enemy. The tactics are alliance, war, marching, encampments, stratagems and recourse to protection. Stratagems literally divides a hostile army into two bodies in order to engage the one while the other is out of position. In Kamandaka Nitisastra, davidhibhava means stratagems which means two-fold, one on his own side and the other on his enemy's side. Strategem is declared to be of two kinds by those who know the sixfold characteristics of tactics. The general commanding a part of the army is sent in one direction and the king is posted in his fortress to defend it with another part. On making alliance and war, the king should employ encampment, marching and recourse to protection after inspection of what is needed. Alliance connected with present or future is of two kinds, i.e. both go together or one goes in one direction and the other in another direction.

To Manu, recourse to protection is said of two kinds: first, when the king is oppressed by enemies and take refuge to save his wealth and second when he is among friends and take refuge in order to forestall danger.[50] The king can take recourse to an alliance whenever a king feels an increase of power in future and at the moment he feels little annoyed. When all the elements of the army are in proper order and when the king thinks himself in a strong position then he should go for battle. He should also march against the enemy when the king is content with his forces and are in prosperous conditions. But, if he is weak in carriage and force then he should encamp strenuously. When the king considers the enemy more powerful then he should avoid battle. But still, then, when the battle is waged and causes injury to the king, he should in that position fearlessly fight against the enemy. When the king has obstructed the enemy, then he should harass his kingdom and even spoil his fodder, food, water and fuel. Manu stresses that water of tanks and wells of the enemy are to be polluted, fortifications are to be destroyed and ditches are to be emptied of their water.[51] These were several method of tactics mentioned by Manu when the king was unable to fight and when methods of open fighting proved unavailing.

The Principle of Morality in War

In ancient India, the emphasis was laid on difference between righteous and unrighteous war. Infact, the underlying idea was to limit battles and confine them to combatants only. Only able soldiers who fought as true combatants could be killed in lawful combat and even in their cases, attention was given to the fact that there the king did not indulge in excessive violence. No attempt were made to increase pains of those soldier who were severely injured or in the process of death. The important question is that are modern wars moral in nature. It is quite debatable.

So far as Manava dharmasamhita is concerned, its precepts constituted not only certain general legal principles or theories but also formed of what may be called religious law whose observance was compulsory. The principles laid down by other authorities had almost the same force and every ksatriya, in ancient time, was required to abide by those rules as far as practicable and regulate his conduct of war in accordance with them. It was felt that battles were inevitable in a dynamic and ever-expanding human society. As such Manu states that it was the sacred duty of a ksatriya to fight against the enemy according to just principles of war.

Thus pre- and post-Manu periods looked upon all warriors to fight battles according to righteous principles. It seems that the ancient kings were much aware of unjust battle or unrighteous warfare. How righteous battle was different from unrighteous it could be seen in the writings of Grotius. Grotius had also laid emphasis on just and unjust battles. He maintained that it was within the perfect right of a state to participate in a just war as opposed to an unjust war. This resembles ancient Indian art of warfare where Hindu jurists laid much emphasis on rational and human rule of warfare and even prescribed certain principles for the regulation of inter-state relations.

The Laws of war in ancient India drew a clear distinction between civilian's and belligerents. While stressing the principle of morality, Manu says that the following persons were not to be killed; one whose is a foot soldiers, i.e. when the other party is seated on a chariot; the impotent, the man with folded hands, i.e. one who is praying for mercy; one whose hairs are disheveled, one who has taken his seat and virtually withdrawn from hostilities, one who has surrendered, one who is asleep, one who has withdrawn from contest, one who is disarmed, one who is averse to war, one who is mere spectator and who is engaged in fighting with another.[52]

Manu, further, says the following are also not to be killed. One whose arms are broken, one who is mourning the loss of his son, one whose body is charged with enemy's arrows, one who is afraid of war and who is reluctant to fight. Furthermore, the use of poisoned weapons were prohibited in the Mahabharata and the Manusmriti. It means that the "use of poisoned arrows, projectiles and peculiarly sharped arrows," were not permissible.[53]

The idea behind was to stop torture and sufferings likely to be caused by these weapons. Even Article 23(e) of the Fourth Hague Convention forbids the employment of arms, projectiles or material of a nature to cause superfluous injury. Dr. Lawrence also asserts that "fighting men may be wounded or slain in wholesale fashion but they may not be tortured."[54]

It is, however, not clear that every rule cited above has been followed in modern war. Today, nations continue in their efforts to excel one another in equipping better instruments of wars. The best example could be seen when atomic bombs were dropped in Hiroshima and Nagasaki in 1945. Even after the end of the second world war with the establishment of United Nation Organisation the objective of banning nuclear weapons have not met with much success. Judged from this viewpoint, the ancient Indian rules on putting ban on the instruments of war would seem to be quite commendable. Large-scale devastation of enemy territory formed a part of unjust war and was never approved in ancient India. Today, modern jurists also have held the view that devastation is permissible as a measure of self-defence. Article 23 of the Hague Regulation declares that it is necessary for any fruitful result in war.

The Theory of Prakitis: Interstate Relations and Diplomacy

The normal state of affairs in ancient India was there were large number of kingdoms and principalities. The maintenance of foreign policy was important for public activity of every kingdom and hence foreign policy served the purpose of its activities with other kingdoms. As Prof. Chatterjee observed that the relations among states in ancient India was taken for granted. It was rather the rule than the exception. So far kingdoms were concerned they could not live in isolation and as intercourse somewhat seems unavoidable.[55] Manu believed that obtaining a friend was more important than by gaining gold and treasury. Even a weak friend, who was righteous, grateful, loyal, was greatly praised.

Hence, Manu expounded the theory of Rajmandala (circle of states). Although the theory of rajmandala is found in Rig Veda but evidences are there to proof that the Vedic Aryans were aware of the rudimentary forms of diplomacy. This is suggested by the description of Agni as messenger when he is represented as "serving in secret passing to and fro between the races and an envoy to mankind.[56] Even in the Artharva Veda Samhita, Agni is given a political bias and was said to be a speedy messenger who was to see the arrival of much tribute. But beyond this, there is nothing to suggest that the Vedic Aryans were aware of a well-developed theory of rajmandala.

When the theory of rajmandala became actually operative is difficult to state but the first clear picture emerges in the Manusmriti. Manu has laid down that on the conduct of madhyama (middlemost), Vijigisu udasina (neutral), and satru (foe). These four constituents (Prakriti) formed the foundation of the circle of neighbours. Along with these, there were eight others which were enumerated in the institutes of polity and hence the total number was twelve. The ministers, kingdom, fortress, treasury and army were five other constituent elements of the circle. Thus, the whole circle consisted of seventy two constituent parts.

We may analyse the concept of Rajmandala as follows:

Circle I –consisting of Vijigisu+ his friend + his friend's friend (three rulers).

Circle II–consisting of the enemy + his friend + his friend's friend (three rulers).

Circle III–consisting of the madhyama king + his friend + his friend's friend (three rulers).

Circle IV–consisting of the udasina or neutral king + his friend + his friend's friend (three rules).

Since each ruler possessed five elements of sovereignty comprising minister, country, fort, treasury and army, each circle had three rules + 15 = 18 elements and, therefore, four circles had 18 × 4 = 72 elements. We, thus, have four primary circles of states, twelve kings, sixty elements of sovereignty and seventy two elements of states.[57] Thus, this theory contemplated a system of a state bounded by hostile, friendly or natural relations with an ambitious potentate as its central figure.[58]

Thus, the Rajmandala theory became the source of foreign policy of the kings. One who was desirous of victory should overcome all the enemies by means of the four expedient, conciliation and the rest employed either singly or conjointly or by bravery and policy alone.[59] The king was supposed to know the secret of his enemies but should conceal his own secret policies from them. He was to cover his secrets and seven elements of kingdom like the tortoise. He was to plan his undertaking (patiently meditating) like a heron; or like a lion, let him put forth his strength: like a wolf, let him snatch his prey; like hare, let him double in retreat.[60] Manu has also referred sarna, ullama and madhyama as three kings but he did not explain those terms.[61] The enemy who was intelligent, was of noble race, was brave, clever, grateful and firm were considered to be the most dangerous enemy. Without neglecting him, he was to make peace with such an enemy.[62]

The word madhyama is explained by Medhatithi as a king whose territory was close to that of Vijigisu. According to Kulluka, madhyama was one whose territory was co-terminus with that of Vijigisu and who was capable of resisting them individually when they were not combined.[63] In this way Vijigisu tried to win over the madhyama. N.N. Law felts that the madhyama was, thus, a stronger power to intermediate between the Vijigisu on the one hand and udasina on the other hand. Both these were states of higher grades of strength.[64]

The udasina was described by Manu as a king who was beyond the enemy and his neighbour.[65] According to Medhatithi such a king is capable of containing each of the two, Vijigisu and his enemy singly but not conjointly and also each of these three Vijigisu, madhyama and udasina singly but not conjointly.[66] To, Kulluka, udasina was capable of assisting Vijigisu and madhyama when they were in alliance and was capable of resisting them singly and when they were not combined.[67] Thus, udasina was brave, compassionate and liberal. He pursued a policy of non-intervention which was acceptable by the Rajmandala.

One of the important functions of Vijigisu was to protect his allies and others against possible enemy attack because their destruction would eventually lead to the ruin of his own kingdom. He was to secure his rear before undertaking an expedition. The parsmgraha (heel–catcher) might be hostile to him and might try to invade his kingdom during his absence.[68]

Thus, a kingdom was secured and protected when its relations with other kingdom were arranged in such a manner that its allies and foes were so interspersed around that at no time it was rendered helpless. Defence to one kingdom was aggression to another.

Six Measures of Rajmandala Theory

The sixfold royal policy was equally related to the concept of Rajamandala. According to U.N. Ghosal, if the former reflected the schematic grouping of states after the pattern of inter-state relations in post-Vedic times, the latter expresses with through adherence to formal rules and the wide range of those relations.[69] Manu enjoined upon the king to mediate upon six measures of royal policy viz., sandhi (alliance); vigraha (diplomatic war); (3) yana (preparedness for war); (4) asana (maintenance of a post against the enemy); (5) dvaidhibhava (double dealing); and (6) samsraya (seeking protection). This policy was to be followed after due consideration of calculating gain against loss.

Sandhi was important for two reasons: one when an army marches together with an ally and second when the allies act separately.[70] According to Medhatithi and Kulluka sandhi meant making present of gold and other things in order to secure the goodwill of both the parties and opposite of this was vigraha.[71] Manu defined vigraha as a diplomatic contest first and then recommended the use of force only for the strong king whose victory might be assumed in war on account of his superior force.[72] Manusmriti classified yana (attack) into two categories viz., (1) to attack the enemy at once in adversity and (2) to take the help of the ally and then attack.[73]

Asana was also of two types: (1) holding the post due to lack of means and (2) holding the post due to the pressure of the allies in face of the enemy.[74] It is prescribed both for weak and strong kings. Medhatithi and Kulluka felt that the policy was meant for ignoring the enemy and also withdrawal of oneself.[75] Davaidhebava was also divided into two categories; (1) to keep the armies in accordance with the wishes of the senapatis at one place in order to meet the end of the kingdom and (2) to occupy some other forts with a fraction of the forces probably to impress on the enemy and the existence of strained relations between the king and the commanders-in-chief.[76] In the last category was samsraya. Manu mentioned different types of samsraya. The first consisted in taking the help of a strong king pressed by the enemy and the second consisted of the spreading of news of the help of a powerful king on the apprehension of trouble from the enemy. The purpose was to check the evil intentions of the enemies.

Fourfold Upayas (Expedients)

According to Manu, the success of the fourfold upayas rested on both fate and human efforts. Fate could not be achieved without human efforts.[77] The king had to accomplish his mission by the proper and timely application of the fourfold expedients of diplomacy. These upayas were sama (conciliation), dana (giving gifts or presents), bheda (causing dissession) and danda (force); These measures are still relevant in the conduct of modern diplomacy.

To Medhatithi and Kulluka sama is friendly meeting, sitting together, conversing, seeing each other's wife and so on. Thus, friendliness would achieve the task of a king. Dana means presentation of gifts inorder to create actual attachment. Kulluka includes elephants, horses, chariot in the list of gifts. Bheda involved all sort of fair and foul means on the part of the king employing it. Manu does not give much importance to bheda.[78] Actually, this was applied only to weaken the enemy, both internally and externally. Kulluka expressed bheda as winning over the subjects and the followers of the king while Medhatithi saw it as winning over the family members of the person desired to be won over. Danda was the last mean of diplomacy. It might be called diplomatic war and not real armed conflict. Rather, it was seen as the last resort to war. Manu allowed use of danda when the other means failed to achieve the desired result.[79] Medhatithi and Kulluka believed that the success was made sure by the use of danda. Manu advised the king to use force slowly and not hurriedly.[80]

These upayas made the king more confident and was able to win over his enemies and influence the activities of madhyama and udasina kings.[81] Manu stated that the king was to work in such a way that his allies, neutral and enemies did not become superior.[82]

The Institution of Ambassadors

In the context of foreign policy, ambassadors played a vital role in the execution and fulfilment of the sixfold policy of the kingdom. An ambassador was called a duta. In ancient India, dutas were appointed for special missions. It was like the modern practice of keeping representatives as permanent agents in foreign states. The function of a duta was to deliver the message correctly, to make or break alliances or treaties, to declare battle or make peace, to study the geographical position and strong points, military strength and financial condition of a foreign kingdom and to gather the greatest possible information. He was, thus, primarily concerned with the vital issues of foreign policy of kingdom. Manusmriti and Kautilya's Arthasatra more or less prescribed the same higher qualification for them such as noble family background, modesty, tactfulness, eloquence of speech, capacity to convey the message exactly as entrusted to him and sharp and excellent memory.

From the Manusmriti one cannot make out what exactly was the status of a ambassador. Manusmriti even does not mention of the salary given to the duta. On the other hand, Kautilya has given a detail account of it. On the importance of dutas, Manu stated that the king had to be careful about the details given by the ambassadors. To a great extent, the existence of the kingdom depended upon these dutas. Manu had also emphasized the sacredness and inviolability of ambassadors. His conduct of behaviour should be good. He should be respected and treated courteously even if he conveyed an unpleasant message because he was merely the mouthpiece of the king. He was never to be killed.

Spies

In India, the system of espionage is as old as Rigveda. Both Manu and Kautilya said that it was a prominent feature of a kingdom. It was regarded as one of the eight limbs of the army. As spies were "the eyes of the kings" they were appointed by the king to collect information about the internal affairs of his kingdom as well as about foreign kingdoms. Hence, they were dispersed all over the kingdom and also spread to foreign kingdoms. If found detected these spies could be ill-treated or even killed by external forces of kingdoms. Hence, spies were to be so clever as not to be detected or identified by others. Manu and Kautilya have given a list of persons who could be entrusted with the risky job, such as braharman, hypocrites, siddhis, person capable of doing impersonation, posing as blind and deaf depending on the situation and should be cunning.

The spies played an important role in diplomacy. It was through them that the king acquired the knowledge of the secret and weak points of enemies. Medatithi also prescribed the same functions of spies. Manusmriti lays down several regulations for the employment of spies in public life. The spies were supposed to know the behaviour of each rural official. Manu enjoined upon the king to be on guard and well armed while meeting the spies.[83] Manu had mentioned five types of spies without naming them. They were appointed by the king to settle the truth in judicial matter. Spies function was to detect thieves who stole property. Among them, the open rogues were burglars, robbers, cheaters, gamblers and who took bribes. These rogues lived by teaching the performance of auspicious ceremonies, sanctimonious, hypocrites, fortune tellers, official of higher ranks, physicians, who acted improperly, prostitutes, men who lived by showing their proficiency in arts and those who were non-Aryans but who walked in disguised like the Aryans.[84] Spies, were also used to constantly ascertain the king's own and his enemy's strength. One will be surprised when Manu does not say about the means to be employed by the spies to create trouble in other kingdoms. Later on, Kautilya mentions of the means to be employed by the spies, either morally or immorally.

Conclusion

So far the history of ancient India is concerned it is largely a history of Hindu culture and progress. We have discussed how the Hindu culture could be defended? It is important to mention there is an intimate relations between means and end. The end was the protection of the kingdom and its subjects and means were the use of force (danda) or army to obtain the desired result. Unlike modern age, army was not developed on mechanical lines but in ancient India much emphasis was laid on courage and individual leadership. Moreover, in the modern age, war has become more complex because of industrialization and invention of new and sophisticated weapons. Today, war and religion are subordinate to politics. But in Manu's period it is assumed that the king enjoyed supreme political and military powers and he had to work according to sacred law. Hence, one can say that since the king was the supreme body, there was an intimate relationship between politics, military and religions. An important question that emerges to what extent Manu's writing is relevant to the modern world? The answer to this question could be seen from the fact that the term military power or force was construed as one of the characteristics of both ancient and modern world. The purpose of army was the protection of society, subjects and kingdom.

Realising the supreme importance of army (military power) the kingdom (small or big) was protected by the king. Not only through the use of force, the king also maintained and preserved the social order, based on varnas and ashramas. As such, Manu has given enough importance to the concept of danda (punishment). This was necessary to instil fear among the subjects so that the society could be safe from evils. This resembles Machiavelli's statement when he says that "where there are good laws there are good arms." As such society was protected from external aggression and internal disturbances. Much before Machiavelli, it seems that Manu thought on similar lines.

Justifying the state as the embodiment of force Robert MacIver, an authority on politics, believes that each state in the ultimate analysis is nothing but the personification of force. The state displays its untrammelled authority in conquest and expansion, in the war of steel and gold and in alliances and diplomacy. On the other hand, Manu's kingdom is quite intelligible in the context of the nature of the modern state. Much have been discussed on Manu's theory of danda. Like in modern state, Manu also regards punishment (danda) inherent in the kingdom. This was not meant so much for the citizens as for the kingdom itself. In support of this, Manu realizes "if punishment is properly inflicted after due consideration, it makes all people happy; but inflicted without consideration, it destroys everything." From this, it can be deduced that the use of force in modern state is sanctioned by the judgment of the society while in the ancient times the ruler's judgment was more important because ancient society, perhaps, did not have the power of judgment of its own. But, it should be made clear that in ancient India no tradition, custom and conduct would have preserved unless these were sanctioned by the sacred laws of society. From this, it is quite clear that force used by the king, as mentioned in the Manusmriti, was conditioned by the will of the society.

Again, there is some difference in the usage of the term protection in ancient and modern times. Manu speaks of the necessity of protection to defend the weak against the strong and it should be within the rules of dharmasastra. But today, protection is related to the minimum standards of living, health and human security. In other words, state should provide social assurance, i.e. in the case of natural or accidental loss the people may be compensated.

Last, but not the least, the term 'defence' and 'means of defence' are vital for every states to maintain its code of honour and promote their national interests. On the ethics of defence, strategic writers believe that force will not disappear from international politics. Today, the debate on moral or immoral principles of war have also assumed much importance. But, it was found that the principle of morality was more followed during Manu's period. Today, capitalism, imperialism, scientific advancements have made war more ferocious. This could be proved from first and second world wars and more recently with the advent of nuclear weapons has made war avoidable because if those nuclear weapons are used it would mean the end of civilization or the society. Still, then today, as it was in Manu's time defence, maintenance of law and order and basic satisfaction of the needs of the people were important for every kingdom developing or developed.

REFERENCES

1. W. Doniger and B.K. Smith, *The Laws of Manu*, Penguin Challics, 1991, p. 16.
2. M.V. Pathwardhan, Manu, Manusmriti, *The Ideal Democratic Republics of Manu*, Motilal Banarsidas, Delhi, Varanasi, Patna, 1968, p. 18.
3. Kelkar, Mahareshtriya, Dnyanakosh, Vol. 23, p. 218.
4. M.V. Pathwardhan, op.cit., p. 194.
5. Manu, 7.29, 65: 10, 1.
6. Ibid., 109-13.
7. J. Basu, India in the Age of the Brahmanas, Calcutta, 1979, p. 86.
8. B.A. Saletore, *Ancient Indian Political Thought and Institutions*, Asia Publishing House, Bombay, Calcutta, New Delhi, Madras, 1963, p.18.
9. Ibid., 61.
10. Ibid.
11. Manu, 8, 417, 7, 35.
12. Ibid., 8, 336.
13. Ibid., 7,110:114; 144.
14. Ibid., 8, 401.
15. Ibid., 8, 47-50; 139-143; 151-155; 158-167.

16. Ibid., 7, 38-39; 79, 82-86; 134, 136; 8, 395; 9, 313.
17. Ibid., 8, 129-30; 310.
18. K. Motwani, *Manu Dharmasastra*, Ganesh and Company, Madras, 1937, p.150.
19. Manu 9, 294.
20. Mahabharat, Santi Parva, Section 68, Verses 8-31.
21. V. Kane, History of Dharmasastra, Vol. 111, 1946, p.18.
22. I. Tope, *The Minister as a Kingmaker*, Kitabistan, 1941, p.52.
23. Manu, 7, 80.
24. V.P. Verma, *Studies in Hindu Political Thought and Its Metaphysical Foundations*, Delhi, 1974, p.240.
25. Manu, 7, 5-9; 9, 303-11.
26. B.A. Saltore, op.cit, p.427.
27. Ibid.
28. Ibid., p.432.
29. Ibid., 436
30. Manu, 7; 208-209
31. See J. Jolly, Hindu Law and Custom, 1928. Narada Legal Procedure III.
32. G. Buhler, *The Laws of Manu*, Banarsidass, Report from Oxford University, Delhi, 1886 ed, 1984.
33. Manu, 7, 2-3.
34. Ibid., 7-8.
35. Ibid., 7,3.
36. Ibid. 5, 96; 7, 4-7.
37. Manu, 9, 253-255.
38. Ibid., 7, 20-34: 8.
39. Ibid., 7, 78,99.
40. Ibid., 8,23.
41. B.A. Saltore, op.cit p.66.
42. E.W. Hopkins (ed), *The Ordinances of Manu*, Oriental Books Publishing, New Delhi, second edition 1971, p.169.
43. Ibid.
44. Ibid., p. 159.
45. BA. Saltore, op.cit. p. 436.
46. Hopkins, op.cit, p.154.
47. Ibid., 173.
48. Ibid., 174.
49. Ibid., 170.
50. Manu, 7-195/196.
51. Ibid., 7-91, 92, 93.
52. Ibid., 7-90.
53. T.J. Lawrence, *The Principles of International Law*, London, p. 529.
54. H.L. Chatterjee, International Law and Inter State Relations in Ancient India, p. 7.
55. B.A. Saltore, op.cit, p. 474.
56. Ibid., p. 476.
57. U.N. Ghoshal, *A History of Ancient Indian Political Ideas*, Delhi, 1966, p. 845.
58. Manu, 7, 159.
59. Ibid., 7, 105-106.
60. Ibid., 7, 87.

61. Ibid., 7, 210.
62. Medhatithi and Kulluka on Manu, p. 7, 155.
63. N.N. Law, Interstate Relations in Ancient India, pp. 12-13.
64. Manu, 7, 158.
65. Medhatithi on Manu, 7, Bhashyam, 1967, p. 158.
66. Kulluka on Manu, 7, Manvarthamuktavali, Bombay, 1946, p. 155.
67. Manu, 7, 207.
68. UN Ghosal, op.cit., p. 94.
69. Manu, 7, 160-163.
70. Medhatithi & Kulluka, op.cit, 7, 160.
71. Manu, 7, 198-200.
72. Ibid., 7, 165.
73. Ibid., 7, 164-66.
74. Medhatithi and Kulluka, op.cit, 7, 160, 166.
75. Manu., 7, 167-168.
76. Ibid. 7, 205.
77. Ibid., 7, 109.
78. Ibid.
79. Ibid., 7, 108.
80. Ibid., 7, 159.
81. Ibid., 7, 177.
82. Ibid., 7, 153, 223.
83. Ibid., 9, 256-60.
84. Ibid., 9, 298.

3

KAUTILYA

Introduction

Generally, it is assumed by the western world that Ancient Indian literature lacked strategic thinking. But when we look back into the ancient Indian history we find heaps of strategic literature that shows intimate relationships between society, state, war and foreign policy. The period under review witnesses a remarkable development in ancient art of war by profound military thinkers. Among them, one bold and intuitive thinker was Kautilya. During his period, society and kingdom under went great social, political and military changes. War was important because it was regarded as a means to obtain social and political goals. Thus, Kautilyan study on politics and war become important not only for factual point of view but also that one can deduce certain principles of war which are of universal nature and can be applied even today.

There was a time when such literature were missing. But the discovery of Kautilya's *Arthasastra* by Dr. R. Shama Sastri in 1905 aroused much interest on the subject. About Arthasastra's composition, it is quite debatable among scholars. Nevertheless, a critical examination of the contents have convinced some scholars that the text was not the work of a particular individual but it belongs to certain schools of thought. They regarded that the work could not be composed in the third century BC. There are scholars like Dr. Ganapati Sastri, Jolly, Dr. Winternitz, Meyers and others who regard that Kautilya should be credited with the dissemination of true knowledge about ancient ideas on state, politics and war. As far as Kautilya's Arthsastra is concerned, it is presumed that it might be written between third and second century BC.

Kautilya, the author of *Arthasastra*, is generally regarded as the contemporary and preceptor of Chandragupta Maurya. Dandin in his Das Kumarcharita believes that Vishnugupta composed a political treatise of 6,000 slokas for the benefit of Mauryan rulers. Banabhatt, describes *Arthasastra* as the science and art of diplomacy. The author of Pancha Tantra assumes that the author of the book was a Brahmin by the name of Chanakya. Ganapati Sastri called him Kautilya because he was of Kautila Gotra and since he was born at Chanka, hence, he was called Chanakya and he was baptised Vishnugupta by his parents.[1]

Arthasastra, the most authoritative and comprehensive work on statecraft, is well known all over the world. In the western world, there are several works of literature on statecraft and political power. In Greece, there were great political philosophers, like Plato and Aristotle, who had written on state and its functions. It would be interesting to study them and compare the development that was witnessed during the Kautilyan time. Aristotle belonged to the greater part of the fourth century BC and Kautilya belonged to the latter part of the same century and to the early part of the third century BC. In modern times, Thomas Hobbe's *Leviathan*, James Harrington's *The Commonwealth of Oceana*, John Locke's *Two Treatises on Government* and Jean-Jacques Rousseau's *Social Contract theory* dealt explicitly with the same subject. Similarly, one could also correlate Kautilaya's book with Karl Marx's *Das Capital*, Joseph Schumpeter's *Capitalism, Socialism and Democracy*, Clausewitz's *On War* and Machiavelli's *Prince and the Art of War*. Hence, for our purpose the comparison between Kautilya, Aristotle and Machiavelli would be an interesting and relevant study because all of them represent two different worlds (ancient and modern). None of them were interested in the ideal state but were more concerned with the security of the state against external threats and internal harmony.

State

Prior to the state, society is important in any period. The protection of society means that people are defended from any kind of threat and this makes society a prosperous one. To protect society is the real art of the state. Kautilya mentions seven elements of sovereignty, which are the primary substance of the state. The term *rajya* occurs in ancient literature but rajya is not a kingdom because kingdom means the dominion of king and the separateness of the king from the dominion is unthinkable.

Kautilya's Arthasastra does not directly deal with the question of the origin of the state and its nature. It also does not enquire how some men come to rule over others and how and why the majority of people are content to be governed by a few. Nor does it refer to the various forms of state's organisation and discusses its relative advantages and disadvantages. Its primary concern was related to practical administration. Kautilya's reference to people suffering from anarchy, as illustrated by "a large fish swallowing a small one first elected Manu to be their king and allotted 1/6 of the grains grown and 1/10 of the merchandise as sovereign dues. Fed up by this payment system, kings took upon themselves the responsibility of maintaining the safety and security of their subjects when the principle of levying just punishments and taxes had been violated. Hence hermits also provided the king with 1/6 of the grains gleaned by them, thinking that it is a tax payable to "him who protect us."[2] Thus, it is laid down that the sale of commodities, whether indigenous or imported, should be arranged by such methods that the subjects are benefitted thereby and also to make sure that any gain which proved to be harmful should be avoided.

What is important for a state? It is the protection and preservation of the lives of its subjects. Earlier, in Manu, we have dealt in detail about various *varnas* and their functions. But Kautilya, further, makes an improvement in its duties. Before him, Sudras were relegated to a life of servility to the three higher classes, particularly to the priestly class. Sudras were placed in an inferior position which was considerably modified by Kautilya. He said that the duty of the Sudras is the serving of the twice born, agriculture and cattle breeding and trade, the profession of artisans and court bards.[3] Kautilya followed the ancient injunction of making the

Sudras perform the duties of the Brahmins, the Ksatriyas and the Vaisyas. While doing so he emancipated them by permitting them to take the profession of agriculture, cattle breeding and trade which was the monopoly of Vaisyas and even adopt the duties of the artisans and of court bards which have considerably raised their social and economic status.[4] Thus, Kautilya had not only raised the social and economic position of the Sudras but by combining all the functions of other classes one can say that this class got certain liberty and in this process helped the state to fulfil its aim, i.e. the well being of society. This element of Kautilya's thought is missing in the writings of Manu's theory of society. But, this aspect of Kautilya's thought should be not taken in a negative sense that Sudras were one independent class. All the classes were inter-related to each other so far the protection of the society was concerned. In other words, no class could survive without the cooperation of the other classes.

Kautilya's state is composed by human element, i.e. people. Similarly Machiavelli also describes the world through human element. The ends, both advocate, are quite similar. The foundation of Machiavelli's political thought is said to be in one line "The Prince": "the wish to acquire is in truth very natural and common and men always do so when they can for this they can be praised not blamed."[5] Although political equilibrium is the ideal a prince should strive for, it remained illusory and "all human things are kept in a perpetual movement, and can never remain stable states naturally either rise or decline.[6] Like Machiavelli, Kautilya also sees beyond the reign of one king to the stability of the realm.

Thus, Kautilya believes that the state without an articulated population is "as useless as a barren cow". He regards the state primarily as an association of human groups. The state is meant for the protection of life and prosperity of the people so that the society can progress. Though Kautilaya mentions the material welfare of the people and the state, the moral and ethical philosophy, which are also essential for the protection of institutions and property are not been considered seriously by him. Similarly, Kautilya's state was not something like Greek's state at that time. To the Greek philosophers, the state was a foreign ideal which assisted not only in the solution of man's ethical problem but also to achieve his highest goal. The individual and the state stood in the closest relationship to each other and the family and the social and religious life of the Greek people and his self-realization were determined by the state. Aristotle conceived of a city state whose territory and population were limited but not thought of an empire. Kautilya, on the other hand, was an unequal champion of a large kingdom and it could grow into a mighty empire. But Aristotle affirmed that the king must have force in order to maintain law. On this, he seems close to Kautilya's view on *danda* which was meant to maintain the state by protecting its subjects. But, the Hindu thought of the past, present and future and the scope of the state's action was not limited. It extended from the solidarity of the narrow social unit of the city state to the humanity.[7]

To Kautilya, rationalism and realism, and not god, was important. Since then, one can say that he was more close to the secular side of life. As such he ascribed the supremacy of ideas and this helped him to renounce the theological background of the origin of kingship. The state, then, is amenable to human control and that the king by Udyama could control human destiny and turn out to be the maker of the age. It also shows that he was not a believer in fate. Like Aristotle, Kautilya emphasized the positive side of politics and gave importance to reason and experience.

Functions of the State

Protection, in ancient times and today, means guarding the state both against internal disturbances as well as foreign aggressions. Of all the ancient theorists, it was Kautilya who was aware of twin dangers that the state can face. It is quite clear from the detailed rules which he formulated regarding the calamities that might hamper the security of the state. Like Machiavelli, Kautilya too felt that a king should endeavour to secure not only his kingdom during his time but even after him. Kautilya even advises that a king should not install on the throne one who is unfit to rule even if he is an only son.[8] Hence, the ultimate end of the king and his government is the health of the state. Thus, like Machiavelli, Kautilya remains loyal to the state and not to the government.

Inspite of giving importance to reason and rationalism over religion, Kautilya, then too, speaks of the protection of dharma of the land within the sphere of state and society. He says "as the duty of a king consists in protecting his people with justice its observance would lead him to heaven. He who does not protect his people or upset the social order wields his royal sceptre (danda) would be in vain."

Neither the protection of the social order nor levying taxes could be possible without promulgating proper laws and maintaining the system of judiciary. In the modern state, the creation of the law has become an essential function of the state. Even in Manusmriti the promulgation of laws had divine tinge but when it comes to Kautilya it was fundamentally a function of the state.

The welfare of the subject held supreme importance to Kautilya. He describes that "in the happiness of his subjects lies the king's happiness: in their good welfare his welfare: whatever pleases him he shall not consider as good but whatever pleases his subjects he shall consider as good."[9]

Explaining further, Kautilya states "of a king, the religious vow is his readiness to action: satisfactory discharge of duties is his performance of sacrifice: equal attention to all is of fees and ablution towards consecration.[10] In a later context, he writes about the supreme importance of this function by stating" strength is power and happiness is the end."[11] While the state was to possess power, its sole object was to promote happiness of its subjects. Kautilya's view about the function of the state in the protection and promotion of the welfare of the subjects have brought quite a relief where no ancient writers on statecraft have done.

The above discussion on the subject could be examined as follows: (1) the relations of state with other states and of the state with the individual and (2) of individual with individuals. Regarding the second one both Kautilya and Manu are silent. On the issue of state's relation with other states, Kautilya has given more detailed information than Manu. Though Manu's talks of Rajmandala theory but it is Kautilya who has elaborated it in such a manner that it looks fresh and interesting. In Manu, there is reference of the relations between the varnas of the mixed classes of the Aryans in general, of father with son and daughter of his band with wife and of the individual with the individual in particular. But Kautilya's discussion on the subject itself form a subject of discussion by itself.

Both the above two aspects aimed at securing order, justice and security could be possible only by the strength of the army. Thus, the ancient Indian state brought into limelight the army, the police and a large number of executive officials. They exercised powerful control and spent huge amount of money for sustaining these institutions so that the primary cause of the subject could be promoted. It is also remarkable that in Kautilyan's work the troops are recruited from all the sections of society. The earlier teachers had favoured the idea of Brahmin army as being the best because of the high spirit possessed by them. Kautilya, while rejecting this view, shows his preference for a well-trained Ksatriya army or a strong Vaisya or Sudra army. There seems little doubt that not all fighters belonged to the Ksatriya varna and that large forces which were recruited came from the Sudra masses and this was quite known then. On the other hand, Manusmriti does not visualize any such possibility.

Depicting the political situation of the state during Kautilyan time one find drastic changes in the system. The Mauryan king maintained an army that could overthrow not only the mighty Nandas but also compelled the most powerful western contemporary rulers who had acquired Punjab from Alexander and made Chandraguta Maurya the undisputed king stretching right from the boundaries of Gandhara to those of Bihar. Hence, in *Arthasastra*, Kautilya not only gives account of administrative and military machinery but also lays stress on the economic resources upon which the Mauryan state could be developed so that the society and the state could be free from internal and external dangers. It also shows that constant combats amongst the Indian kingdoms brought in Kautilya's mind the necessity of a strong and unified state and centrality of force in politics. Here, both Kautilya and Machiavelli drew the same conclusion. What is important is the universal applicability according to their law. Gilbert comments on Machiavelli, "The Prince and the Discourses were intended to reveal the laws which govern the world of politics."[12] On the other hand L.N. Rangarajan regards Kautilya "as a great preceptor of statecraft whose teaching have a universal validity.[13]

The pertinent issue, then, arises if the state is so concerned with the protection of people this supposition could lead to an idea of welfare state. It is supported by Kautilya when he says of those people who have been necessarily maintained by the state and the duty of the state is to maintain particularly minors, aged and distressed people then one can imagine that

Arthasastra is no "police state" nor a merely "tax gathering state" but more a "welfare state".

Limbs of the State (Saptany Theory)

The limbs of the state are called *prakritis* (elements) of the body politics. There are seven in number: (1) swami (king); (2) amatya (minister); (3) janpada (territory); (4) durg (Fort); (5) kosa (treasury); (6) bala (army) and (7) mitra (allies). This is known as the saptang theory. The theory occupied an important position in ancient Indian military thinking and formed the core of the ancient view of the state. It implied that the state possessed organic and concentric entities.

Swami (King)

The king occupied the central position to the state prosperity and defence. Kautilya went to the extent by proclaiming that the "king was the state".[14] The king was to the saptangs as the brain is to the human body.[15] Ishwar Topa brought out the true importance of the statement of Kautilya in the following words. "Its kingship goal was the goal of the state. Kingship was the spirit and the state was its expression. It embodied the ideas of the state. One without the other was inconceivable. The life of kingship was the state. Both were inseparable. Coexistence was their law. The welfare of the state depended upon the kingship."[16]

Kautilya quotes another example of the central position of the king in the affairs of the state. "As the king so his people". He believed that the kingship could change the cultural life of the subjects according to the virtues or vices. This phenomenon is also witnessed in present today statecraft. Hence, he rightly points out by saying that society becomes corrupt when the people at the top, the ministers and the legislature indulges in corrupt activities.

The king, to Kautilya, was the source of social progress and prosperity and everything depended upon him. He appointed the ministers who were responsible to him and held the office during his pleasures. He also appointed *senapati* of the armed forces as well as the heads of the various administrative departments, judges, the collector of revenues and so on. More significant the king wielded the rod of danda. Kautilya looked upon the king as the supreme protector of the people's so that he could bring stability in society. In this sense, the king was rightly called the highest and the most important organ of the state because he was the prime cause of unity. Unity was an indispensable feature of the state and had to be retained at all cost. It was generally assumed that if the king was virtuous, the state prospered. Therefore, a good king was the cause of all progress and bore all the burden of his subject. As such, the king was responsible for the smooth functioning of government's machinery.

Amatya (Minister)

Amatya was the second limb of the state. In Arthasastra, it appears that the amatya constitutes a regular cadre of services from which all high officers were recruited, excluding the mantrins. It is said that, "having divided the spheres of their powers and having taken into consideration the time and place and the work they had to do, all the persons should be appointed as amatyas but not as mantrins".[17] From this, it appears that Kautilya regarded mantrins, as compared to amatyas, higher in status and grades. In other words, the term amatya meant higher officers whose work was to manage the affairs of the various departments and in this management they assisted the king. The mantrins, then would mean the ministers or the mantriparishads.

Furthermore, Kautilya clarifies that it was difficult for the king alone to rule efficiently and successfully. So he has to rely on them as well as to see number comprising the council of ministers. Different scholars have different views about the size of the council of minister. Manu believed that their number should be twelve. On the other hand, Kautilya, showing his practical and administrative experiences, left its exact number to be decided upon the work and requirement of time. Realising the secrecy of decision taken, Kautilya, like Manu, is in favour of three or four members, instead of all members, for consultation related to the governance of the state. He also advocated the idea of setting up of an inner cabinet consisting of few trusted and experienced ministers. According to Kautilya, amatya should belong to the state. He should be farsighted, wise, of good memory, born in high family, influential, well trained in arts, vigilant, eloquent, bold, intelligent, full of enthusiasm, strength and endurance.

Janpada (Territory)

The third limb was janpada. It constitutes two elements: land and population. In modern times, the term geography is used in the context as one of the elements of state or national power. Kautilya advised the king to include people from other countries to migrate and settle in new villages on old or new sites. They could also be brought from thickly populated areas of his own kingdom to settle down in such villages with a view to protect the village and in that case each village should consist of not less than 100 and not more than 500 families. It should also consist of sufficient number of sudras cultivators.

According to Kautilya, the king shall make provision for grassy grounds on uncultivable major parts of land. Brahmins will be given forest land for plantations, for religious learning and for the performance of acknowledgment of wrongdoing. Manufacturers will set up to prepare commodities from forest tract. The superintendent of forests with his association will not only maintain the up keep of the forest but will also make himself aware of all passages for entrance into or exist from such of them as are mountainous or boggy or contain rivers or lakes.[18]

Durg (Fort)

For the defence of the state against enemies, Kautilya states that the king would construct forts on all the four sides of border of his kingdom and he would also establish main cities and capital in middle of kingdom. There is nothing corresponding to the western scheme. The reason for the construction of forts were due to the geographical situation in northern India and the nature of fighting in ancient times. Kautilya mentions four different types of forts:

1. Audaka (water) fort which acts as an island and is situated in the centre of a river.
2. Parvata (Mountain) fort which acts as a rocky tract or a cave.
3. Dhanvana (Dessert) fort which acts as a wild area lacking of water and overgrown with thickets growing in poor soil.
4. Van (Forest) durg which is full of wagtail, water and thicket from four sides.

While Kautilya follows some of the ideas relating to the location of the forts as mentioned by Manu, he has also made his own contribution. Of these fortifications, the water and mountain forts were best suited to defend popular centre. Desert and the forests forts were proper for protection of a king in difficult situations. The king would place weapons in forts.[19]

Kautilya gives a detailed account of the fortified capital which was to be the seat of the royal treasury. It was to have three ditches, ramparts, parapets, roads, towers, passage for exit, passage for the movement of the troops, an entrance gate, turret, canals, pools and defensive weapons including explosives, machines, etc.[20] He also mentions with in detail about various types of building that were to be developed inside the fort. Kautilya was fully aware of the needs of developing a great capital, which even in its ruined state seeks to attract attention.

Kautilya also showed the importance of the relationship between the fort and the people. To him, people were important. As far as fort and treasury were concerned, he gave importance to the fort. Hence Kautilya laid emphasis on people and makes it clear by stating which is better of the two, "the tract of land with forts or that which is thickly populated? The latter is better; for that which thickly populated is a kingdom in all the senses. What can be depopulated country like a barren cow be productive off?"[21]

Thus, we see how he had made his concept more pragmatic and practical. It is quite evident that he had definitely made much progress since the days of Manu. As such, it would be difficult to accept Prof. Aiyangar's view that Kautilya attached scant attention to forts. Equally important is the relevance of forts in the present time. The concept declined with the industrial revolution and with the development of more and more deadly weapons. These weapons made forts vulnerable. But one should remember that even in the nuclear age if nuclear weapon are used it could mean the end of humanity. Inspite of this danger, strategic experts talk of nuclear bunker, which can protect the people. But the construction of these bunkers are far more costly than the forts that were build in ancient and medieval ages.

Kosa (Treasury)

Both Manu and Kautilya's views were quite similar on *kosa*. It was one of the essential functions of the king to guard his finances and keep the kosa full. The principles source of revenue of a kingdom in ancient India was taxation. It was the duty of the collector general of revenue to collect tax on forts, mines, building and gardens. He could increase or decrease expenditure. The collector general of revenue would collect tax in seven ways.[23] This was called "body of tax".

1. Fort – It included tolls, fire, weight measures, the town clerk, the superintendent of coinage, the superintendent of seals and passports, the prostitute, gambling, building sites, the corporations of artisans and handicraftsmen, the superintendent of goods and taxes collected at the gates and from the people.
2. Country – It included sixth portion of the crops, religious taxes, taxes paid in money, merchants, superintendent of rivers, ferries, boats, ships, towns, road cross and ropes.
3. Mines – It included gold, silver, diamonds, gems, pearls, corals, shells, metal, salts and other minerals.
4. Setic – It included flower gardens, fruit gardens, vegetable gardens and field crops which are grown by sowing roots for seeds.
5. Forest – It included game forest, timber forest and elephant forest etc.
6. Herds – It included cow, buffalos, goats, sheep, asses, camels, horses and mules.
7. Road of traffic – It included land and water ways.

Besides this, Kautilya even permitted the king to take even 1/3 or ¼ of taxes in times of emergency or natural disaster. Whenever he increased taxes he had to obtain the consent of the people by appealing to them. Another principle was that the taxes should be neither heavy nor too less. They should be moderate. The king was advised to follow the example of the bees or the calf. Just as the bees collect honey from flowers but without damaging them in any way and the calf drinks milk without in any way bruising the udders of the mother cow, in the same way the king was expected to tax the people without harming or overloading them. The king was advised, never to forget, that nothing exposed him to greater danger than oppressive taxation.

It should be noted that Machiavelli holds a different view on economic power. Machiavelli's thoughts are radically different from Kautilya. He emphatically states, "there cannot, therefore, be a more erroneous opinion than that money is the sinews of war."[22] Machiavelli's argument is that one cannot wage war with gold but needs iron. Without the strong spirit of goods soldiers, an excess of gold merely encourages neighbouring states to plunder the prince's realm. Money is a necessity, Machiavelli argues, but a secondary one that good soldiers could overcome. But to Kautilya, kosa was one of the essential limbs of the state.

Bala (Army)

The sixth limb of the state was bala (army). Some writers called it danda. But bala was more appropriate than danda because danda included punishment. In ancient times, bala was connected with forts and therefore, the major objective of bala was to defend the kingdom against aggression. In Rig Veda there are references that the Vedic Kings were to defend their territories as well to carry war into the enemy's territories. The idea of entrusting the duty of defending the kingdom was given to a caste made of warriors. Kautilya's idea of the different kinds of bala was unlike that of earlier times. In Manusmriti, it is stated that the king should inspect the armed forces daily. But Kautilya improved upon it by introducing a new idea. On his routine checkup, he states that the king attired in military dress and having mounted on a horse or a chariot or an elephant shall go to see his army drawn up in military array.[23]

It would be interesting to compare the views of Kautilya with ancient Greek writers. Plato believed that the legislator should focus in two arts: the people and the territory. But Aristotle went one step ahead and wrote that the legislators should not forget the neighbouring countries and that "a state must have such a military force as well survivable against the neighbours and not merely useful at homes." Furthermore, Aristotle commented on the need to have wealth, freedom, justice and valour. Military power is essential and without it a state cannot, survive.

About the various kinds of armies, Kautilya regarded regular or standing army as the best of all. Let us consider the following statement of Kautilya: "Coming down directly from father and grandfather (of the king) ever strong, obedient, happy in keeping their sons and wives well contented, not averse to making a long sojourn ever and everyone invincible, endowed with the power of endurance, trained in fighting various kinds of battles, skillful in handling various forms of weapons, ready to share in the meal to woe of the king and consequently not falling foul of him and purely composed of ksatriya caste it is the best army." Kautilya did not mind recruiting soldiers from the Vaishya and Sudra classes but did not favour a contingent of Brahmins because they were more prone to be influenced by the supplications and prayers of the enemy. Hired or mercenary troops, organized by co-operations or srenis, troops belonging to the friendly states, troops consisting of deserters from the enemy and those recruited from wild tribes constituted the other kinds of army. Kautilya divided his army into divisions, and subdivisions, each with its own flag, trumpets, drums, etc. He also provided physicians and surgeons for those on the battlefield.

It is interesting to note that Kautilya has given emphasis on employing mercenaries and auxiliary forces. Here, Machiavelli holds a different view and wishes that the prince should not rely on mercenaries or auxiliaries. He says, "mercenaries and auxiliaries are useless and dangerous; and if one holds his state based on these arms, he will stand neither firm nor safe: for they are disunited, ambitious and without disciplines, unfaithful, valiant before friends, coward towards enemies: they have neither the fear of God nor fidelity to men and destruction is deferred only so long as the attack is; for in peace one is robbed by them and in war by the enemy. The fact is, that they have no other attraction or reason for keeping the field than a trifle of stipend, which is not sufficient to make them willing to die for you...no principality is secure without having its own forces."[24]

Mitra (Ally)

The seventh and the last limb of the state was *mitra* (ally). The inclusion of mitra as an element of state by the ancient Hindu writers highlight the actual nature of their approach to the study of politics and war. From them, the state was a concrete entity and had strong affinity with other entities of the same nature. It was one political entity where it was not only confined to internal factor but also affected the external factor. It was something dynamic and changeable according to the prevailing environment and time.

In the course of his treatment towards mitra, Kautilya was too focused on the doctrine of Mandala. The doctrine constituted the basis of external sovereignty. The Mandala doctrine underlying as it did the Hindu ideas of balance of power pervaded the entire speculation on the entire subject of foreign relations.[25] The doctrine envisages the existence of a large number of kingdoms where the relations amongst them were always peaceful and friendly. On the other hand, each was quite apprehensive of its neighbour that it would expand its territories at any cost. Today, this state of affair is not present in India because India is a federal state with a unitary character. On the whole, the doctrine of the Mandala revolved around the king, known as Vijigisu. It was therefore, the main aim of the state was the preservation of internal peace and order by wielding the rod of danda or punishment.

Further analysis of Kautilya military thinking leads to the reexamination of the term war and the role of army in safeguarding the territories and promoting peace and harmony in the kingdom.

On War

Kautilya's rise was due to the prevailing environment where he had to face much humiliation of his time. He actually lived in a period of constant warfare. He soon realised the importance of war as an important aspect of statecraft. It is said that Kautilya was the cause of Mauryan victory by destroying the Nanda power. Victory, to him, could be possible only through the successful conduct of war.

Kautilya mention three kinds of war: prakashyudh (open war), kutayudha (concealed war), and tusniyudh (silent war).

Prakashyudh

Prakashyudh indicated time and place. Prakashyudh was also called dharmayudha because it was fought on the principles of dharma. There was nothing secretive about it. Preparations for such a war were made openly in the full knowledge of the enemy. According to Kautilya, a king whose military forces were powerful and was successful in diplomacy and all kinds of fears and challenges were dealt firmly then he should go for prakashyudh. He introduced a political term which was quite innovative in his time. That was dharmavijaya, which was perhaps, equivalent to dharmayudha. Accordingly, the king was to be satisfied with the acknowledgement of the overlordship by the conquered or defeated kingdoms. The reason, as articulated by Kautilya, was that indulging in such type of war the king should avoid war and bloodshed at all costs as it would promote peaceful and diplomatic relations with the people of the kingdoms and foreigners.

It is the contention of some scholars that dharmavijaya was essentially a term of enormous military significance and that when Ashoka, born and bred in the Kautilya school of politics, employed this term in his thirteenth Rock Edict, he did not mean conquest by morality, for conquest by arms actually transformed him when he saw the horror and the plight of the people killed in the Kalinga war. Ashoka would be at peace with the whole universe and adopt only when the righteous method of warfare was inevitable.[26]

Kutayudha

The unrighteous war in which principles of dharma were not considered was regarded as kutayudha. In Atharva Veda we come across the term kuta. The term kuta, in the context of hunting, ensnaring, was used for a trap or snare. Consequently, in the context of warfare it means trapping the enemy. According to Brahmanda Purana, Kutayudh was also known as Citrayudh in which maya was put into practice. It was said that a certain commander of Bhanda resorted to this and created a Sarpani from whose limbs thousands of poisonous snakes came out in torrents and bewildered the saktis on the other side.[27]

Thus, Kautilya is clear in his idea when a king failed to indulge in prakashyudh he should, then, go for Kutayudha to defeat his enemy. It is an acknowledgement of an unrighteous war and the sacrifice of all the principles of morality. Kautilya, in the context of kutayudh, also mentions the term lobhavijaya and asuravijaya. It was a method of warfare which permitted the use of deadly weapons against the enemy, often in the form of intrigue, poison, charms and spells; necromancy, magic and sorcery, incendiarism and incantations.[28]

Kautilya was an able exponent of Vedic ritual and of Atharva Veda which accepts this method of war only as a defence and of retaliation against the enemy of the king. For the king, when faced with any kind of difficulty, he should rise to any occasion and meet even the challenge of cunning and intrigue by similar methods. It included stirring rebellions in the enemy country, by attack at night, surprises, devastations murder's and assassinations and practice of maya and indra jala. All these were included in the category of kutayudha.

As Kautilya gives details of kutayudha, there is a misapprehension that unrighteous and irreligious wars had his sanction and approval. But as a military thinker, Kautilya had to examine every aspect of human policy including human pugnacity and war.[29]

Tusniyudha

In Tusniyudha, secrecy was of paramount importance. The king would prevail only by maintaining secrecy when striking again and again. The method adopted in this kind of war were similar to that used in Kutayudha. This kind of war was different from prakashyudha as it strikes the enemy silently. The principles underlying this method of warfare are mentioned in Rig Veda and Atharva Veda where number of battle charms with the use of plants like asvattha were used to defeat the enemy.[30]

Thus, it is quite evident from the above analysis, Kautilya had to refer every aspect of war while writing treatise on war polity. To an impartial student of war and politics, it seems that the ancient Hindus had evolved precepts on fair fighting, which formed a courageous code of military honour. There is also no denying that these laws of war were so varied and complex that it led to different kinds of wars, relying more on immoral principles. Later on, by applying the indifferent principles of war led to the defeat of Grahavarman and Prithvi Raj Chauhan.

Kautilya's deep analysis of war leads to other aspects of war. Kautilya was much more concerned about the prevailing conditions of his time.

War or military forces led to the proper formation of the army. What organization was at that time, Kautilya demonstrates a vivid picture of it.

Organisation of the Army

Chandragupta Maurya had a large standing army and efficient war office to supervise it. Generally the army was divided into four divisions. These were patti or padati (infantry), asvi (cavalry), ratha (chariots) and last (elephants). These various components of army were called the chaturangabala. Each division was supervised by their respective superintendents. In other words, Kautilya had also recognized the importance of these well-organized military departments. The functions of these superintendents of various department are mentioned below:

Ayudhagaradhyaksa (Superintendent of Armoury)

The duty of senapati was to collect all sorts of weapons, defensive and offensive, engines of war and keep them in proper order. This superintendent, Kautilya says, "shall employ experienced workmen of tried ability to manufacture in given time and for fixed wages, wheels, nails, armour and other accessory instruments for use in battles, in the construction and defence of forts or in destroying cities and stronghold of enemies."[31]

About weapons, three main types are prescribed which was actually maintained by the armoury department. The fourth category consisted of battlefield weapons such as bows and arrow, spears, swords, daggers, shield, etc. The second category comprised those weapons for defence of the forts such as stones, catapults and bows and arrows. In the third category were those weapons which were meant for attacking enemy fortifications. It included scaling equipment as well as flaming arrows and other incendiary weapons. This shows that weapons in ancient time, like modern age, were of defensive and offensive nature which proved destructive to gain any kind of war goal.

Asvadhyaksha (Superintendent of Horses)

The superintendent's work was to register the total number of horses as received in the form of gifts, purchased for war and also received in return for help. It should be stipulated in a treaty or temporarily borrowed according to their pedigree, age, colour, marks, class and source. The horse attendant received from the treasury a month's allowance and looked after them. He also saw that those horses which were incapacitated for work by war, diseased, and aged would get food for maintenance. Horses unfit for use in war were to be used as stallion (to breed) on mares in the interests of the citizens and kingdom.[32]

Kautilya says that the best horses which were fit came from Kamboja, Sindhu, Aratta and Uanayu, the middle one was from Bahlika, Papeya, Sauvisn and Titali and the rest were inferior in nature. Overall, Kautilya had emphasised that all sided work of a horse was connected with war.[33]

Hastyadhyaksa (Superintendent of Elephants)

Kautilya classified them according to work. These elephants were used for training, war riding and then the last rogue one. The war elephant had seven ways of doing works; standing in attendance, going round, arching together, killing and trampling, fighting with elephants, assaulting towns and fighting in battles.[34]

Rathadyaksa (Superintendent of Chariots)

The duty of the superintendent of chariots was similar to that of superintendent of horses. He should establish factories for the manufacture of chariots. He speaks of various uses of chariots such as temple chariot, festive chariot, war chariot, travelling and carriage, chariot for marching against an enemy's city and chariot for training.[35] He should be conversant with arrangements concerning bows, striking weapons, armour and accoutrements and the employment of charioteers, chariot attendants and chariot horses in various works.[36]

Pattyadhyaksha (Superintendent of Infantry)

The superintendent should be conversant with the strength or weaknesses of hereditary, hired, banded, allied, alien and foreign troops with military operations in water or in high grounds, with open or tactical fighting in trenches or in the open by day or by night and with the employment or absence of employment in different types of works.[37]

The commandant of the army were trained in the science of all kinds of fights and weapons and were renowned for riding on elephants, horses or in chariots. Thus, one sees that Kautilya has given sufficient attention to various military departments and their assigned works inorder to protect the kingdom and make the people happy and prosperous.

Composition of Army

Kautilya assessed the six types of troops which could be made available to the king as and when required. These were maula (standing army), bhtra (local volunteer/auxiliaries), sreni (mercenaries), mitra (troops of ally), amitra (enemy deserters), and atavi (tribal levies).

Maula

Maula means standing army. Kautilya was in favour for having such a army for a victorious king. This army was composed of soldier's who have served the king's family for several generations. The inherent qualities of this type of army was that the soldiers were loyal, disciplined, well trained and well equipped for war with best weapons. It is, however, only prudent that a portion of this force was left for the protection of the kingdom. The rest were meant for engagement in a campaign. This kind of force also formed a large part of expeditionary force if the enemy's force were well equipped and trained. In case the campaign proved to be difficult then the surplus maula troops met the requirements of the capital and the rest of the kingdom.

Bhrta

In this, volunteers formed as a part of the troop. The objective was to engage them in battle. It included veterans and first time volunteers who were usually trained in the handling of the personal weapons. Basically, they were from agricultural and small trader sectors. As natives, they were concerned with the security and welfare of the state. Such kind of troops were reverted back to their profession once the goal was attained.

Sreni

These forces were well-trained and organised bodies of mercenaries under there own leaders who were brought into the battlefield for some specific objectives. Their employment was recommended when the enemy's forces comprised of mercenary forces too. It was anticipated that a king could not solely rely on them inspite of their large numbers, as they avoided hard fighting. Though Kautilya supported this kind of force Machiavelli in his time was against recruiting such a force.

Mitra

These troops consisted of foreign forces. Their proper utilization could be made possible when they were present in large numbers in time of crisis. Kautilya believed they would be useful for short campaigns because of chances of early victory was within the sight of the king. Again, unlike Kautilya, Machiavelli did not favour such types of force.

Atavi

These forces comprised of tribesmen from the forests. They joined the king's army but they had their own commander. The main objective in battle was to collect loot. They were unreliable and as dangerous as a snake. They were employed because of following reasons: (i) to attack the enemy's cities and irregular troops; (ii) to delay the employment of the main force; and (iii) it was politically expedient to remove them because of disloyalty towards the king.

Thus, Kautilya held the view that an army composed from different sources and were capable of fighting could be regarded as an energetic army. Similarly, force belonging to the same region, caste or profession could be mighty army because they would indulge in battles or campaigns if the kingdom is short of food. They were brave and loyal to the king.

Units of the Army

According to Kautilya, a squad of ten under the leadership of an officer was called Padika. It formed the smallest unit of the army. "For every ten members of the constituents of the army, there must be one commander called Padika, ten Padika under a Senapati, ten senapatis under a nayak.[38]

The above gradation closely resembled Shivaji's infantry which was carefully divided into regiments, brigades and divisions. The smallest unit consisted of nine foot soldiers and the officer commanding it was called Naik. The Haveldar of the infantry had 5 units under him. Over 2 or 3 haveldars were placed jamaldar. The officers commanding 10 jumlas was called Hazari and the Sarnbat had 10 Hazaris under him. He was the supreme military and civil head of the state next to the king.

Thus, the composition of the army as suggested by Kautilya was, thus, a mixed one with all its inherent merits and demerits. The age of Arthasastra was an age of imperialist state and an imperialist state believed on conquest and annexation.

Battle Array

Kautilya, therefore, discusses about the conventional warfare of its time. It prescribed standard battle arrays (Vyuha) which have a centre, two flanks and two wings. Each component of the Vyuha was conceived as being of equal strength containing between 9 and 21 units. Each unit in turn, should be based on elephant or a chariot with 5 horsemen and 15 infantrymen in front and rear.

Great emphasis was also placed on reserves behind every battle array. This was the position of the king where he stationed himself. The Arthasastra has shown preference for mountain or forts to station the reserves. With the reserve force, there would be physicians and medicaments to treat the wounded and field kitchens were run by women. The women were also trusted with the task of encouraging the troops.

Selection of Sites

Before camping it was essential for the army to select the proper site for fighting. The state was inspected by the pioneer regiment. Usually a ground near a forest or river was preferred. The Pandu camp in Mahabharata and Harsha's camp were pitched respectively along the Ajiravati river. Even in Mahabharata, a ground level which is cool and full of grass was recommended for this purpose. Kautilya, however, gives a detailed description and suggests specific types of land for each constituent. For entrenching horses, for example, the best ground was a level plain with no rocks and mud but having a number of trees scattered all over the areas.

Sivira (Camp)

Camp was regarded as an important strategical point by the commander. Camps were places where people stayed when they were away from the homes. The armies were also required to stay in them for a break after the day's fight or for the night's rest. Kautilya mentions of military encampment of a king when he is out for an expedition within his own kingdom or in the dominions of another king where soldiers and officers prepared themselves for fighting. The Brahmavaivarta Purana describes it as a camp having ditches with girth of high walls and furnished with 12 gates surmounted by lion archers. Skandhavara was the soldier's camp near the vicinity of the sivir. Senamukha was a military base or a rear station from where the military staff helped their brothers.

Honest and loyal men were posted to watch the movements of the army. Outside the camp and at a distance from it, forest men were posted in circular arrays. It is recommended that other available camping grounds should be such as to place the enemy at a disadvantage. In the camps drinking, gambling and other social enjoyments were prohibited and for men coming in and going out of the camp a system of passports[39] was maintained inorder to distinguish them from the enemy's spies who might enter in disguise and unobserved. The camp was seen as a self-sufficient unit. Thus, Kautilya says that "the king and allies inorder to secure their triumph should observe in their encampment the strict rules of self-denial, liberality and religion".

March

Before the march outbreak of battle, the king addressed his forces and emphasised that he was one of them. Next the senapati addressed them and announced rewards for acts of bravery. Whatever loot the soldiers captured would be theirs but senapati also announced gratuities at the end of war.

Once all this was done and the king had finally decided to attack, the movement of the armed forces was the next step. The first thing a king did was to collect geographical data and the route where the enemy was placed. The next consideration is the strength of the enemy. According to Herodotus, Darius used to get the route of the river Indus surveyed through his commander Skylax before attacking the enemy. Alexander did the same work through his commander Nearks. The Niti works also advise the king to ascertain the nature of the country, the type of his force before fighting. More or less similar views have been found in the writings of Manu, Visnupuran and Kamandaka.

According to Kautilya, actual march was seen in terms of short and long halts. During the march, the army should be well provided with commissariat, supplies, transport and communication facilities. Due consideration was also given to time factor. Recognising the importance of time and place Kautilya mentions: In day time, the crow kills the owl. At night, the owl kills the crow. "Similarly a dog in land drags a crocodile and a crocodile in water drags a dog. This shows the place of the fight is important." Further, it seems that the month of December was generally regarded for long marches and March–April for short distances. Kautilya says that during hot season only asses, camels and horses were to be used in preference to elephants. The dewy season alone was fit for the use of elephants.

The nayak led the army while the senapati guarded the rear. If the route of the march was sufficiently broad, it was recommended that the front ranks should march in a crocodile-like order (makara), and the rear in a cart like order, both flanks should march in a diamond like array. But, if the path was a narrow one, the army was to move in pin-like order (suci). These kinds of marches in modern times are explained as perpendicular and parallel or flanks marches. The march was undertaken only after the roads had been examined and cleared of thorns and shrubs.[40]

Even on the march, the army was arranged in battle order. The purpose was to face the sudden attack. If the enemy creates hurdles by throwing missiles from the other sides, the invaders should withdraw from there and move to another favourable point to cross without the enemy's knowledge. Attentions were also given to the base of operations because much success depended on the communication with army stationed in bases. A river or mountain fortress was regarded as the best base.[41]

Whatever be the outcome of the battle, the Arathasastra is scrupulous about one principle. After the battle is over, the army should not cause harm to people of the enemies' king. Even, when capturing a fort, the people inside must be assured of their safety and be allowed to leave the fort for safe places. If the territory was annexed only the king was forced to become an ally or a vassal and the people were to be won over through all means. If the king was reduced to vassalage, he still controls the territory and the army was not obliged to help his king militarily. Their customs and goods must be respected by the new king. After the war, loot was forbidden.

Inter-State Relations and Diplomacy

In ancient India the state was divided into various kingdoms and principalities. Every kingdom maintained foreign relations with not only other kingdoms but went beyond by establishing the relationship with external powers. For example, Seleucos Nicator sent Megasthenes as an ambassador to the court of Chandragupta Maurya. Deimachus and Dionysius were attached to the court of Bindusara. Ashoka's edicts show that great kings had established relations not only with Ceylon but also with state like Syria, Egypt, Macedonia, Gyrena and Epirus. Prof. Rapson points out that these ambassadors were typical of a class. It is in every way probable that constant relations were maintained between India and the west during the Maurya empire.[42]

Kautilya establishes an interesting relationship between war, foreign policy and diplomacy. His major thrust was on the king becoming a world conqueror (Vijigisu). Thus, his emphasis was that the foreign relations depended on making friends, identifying enemies, making treaties, using silent war or war of assassination against and unsuspecting king sowing discord among the enemies, using religious and superstition in demoralising the enemy forces, the spreading of disinformation and his human treatment of conquered soldiers and subjects.

Kautilya vividly describes this as the world conqueror or cakravartin who was not conquered but conquered regions beyond the borders of India. Cakra means wheel. It was possible that the Indian concept of the world conqueror involved someone who ruled as his chariots would roll without obstacles. In this sense, Dikshitar is correct when he says that the idea of world conqueror in ancient India led to an "imperialism" that was "one of the causes of chronic warfare.[43]

By developing the interstate theory Kautilya has emphasised that it is natural that states interact with each other through dissension and force. As a realist thinker, Kautilya argued that there would be conflict among states and as the result the strongest will rule. He claimed that one who knew his science of politics could conquer the world, that one possesses of personal qualities, though ruling over a small territory conversant with the science of politics, did conquer the entire world, could never lose.[44]

Kautilya assessed that by doing this he has not lost anything. A king who carried his duties, ruled according to law, meted out only just punishment, applied the law equally to his son and his enemy and protected his subjects not only went to heaven but had conquered the world upto its four ends.[45]

In order to determine the kind of policy to be adopted in each case, foreign rulers were classified by Kautilya under four heads: ari (enemy), mitra (friend), madhyama (mediator) and udasina (neutral). Inimitable and friendly rulers were classified as natural and artificial rulers. A king and his immediate neighbours, were, according to Kautilya, natural enemies to each other. Abul Fazal, while describing the Hindu system of public administration, said that "the prince whose territory adjoined to his although he might be friendly in appearance yet ought not to be trusted". He was always prepared to oppose any sudden attack from that quarter. A king, who attempted to give trouble to another king without reasonable cause was an artificial enemy of that king.[46]

The ruler whose territory was separated from that of another ruler by the territory of an enemy and whose friendship had come down from father and grandfather was a natural friend. The best kind of friend, according to Kautilya, was he who was constant, noble, straight forward and whose friendship had been inherited from father and grand father.[47] A ruler whose friendship was counted for the sake of the protection of life and property was an acquired friend. The ruler whose territory was situated between the territories of two rival kings and who was powerful enough to help or resist either of them or a mediatory king, was a neutral friend. The distinction between neutral and a mediatory king was not clear. Perhaps, the term udasina (neutral) was applied to a king who remained passive in regard to both the contending parties. The madhyama king was one who exerted his influence to bring about reconciliation. The third and fifth states were likely to be inimical to the king. If the madhyama king is on good terms with both these classes of states, a ruler had to be friendly with him: otherwise he was to ally himself with the second class of states.[48]

The neighbouring kings belonged to one or other of these four classes, namely, parshnigraha (rearward enemy), akranda (rearward friend), parshnigrahasara (ally of a rearward enemy) and akrandasara (ally of rearward friend).[49] The king would make an alliance but would not maintain relations who was far remote. A great deal about the circle of states had been dealt in the chapter Manu. Still then, it is important to list that a circle consisted of three kings, a ruler, his friend and his friend's friend. As each of these kings was supposed to possess six elements of states, namely, the king, amatya, janpada fort, kosa and army. Thus, a circle considered of states, 12 kings and 72 elements of states.[50] A powerful and wise king had always tried to make himself the centre (nabhi) of the circle and to make the friendly powers the spokes of the wheel (nemi)[51].

Diplomacy

According to Kautilya, for a state, political, economic and military interests were important. Diplomacy depended on these interests of the state. He believed that the position of kings were one such that they were either at war or preparing for war. Thus, diplomacy was another weapon used in constant warfare. To him, diplomacy was a series of actions taken by a state, which eventually strengthened the position of the ruler where diplomatic efforts were made to stabilize and protect its interests. He believed that the treaties should be made in such a way that the king benefits and serves the interest of the kingdom. He also made reference of violating treaties and creating dissension between states so that his kingdom might benefit. This was quite similar to Bismark's strategy. In fact, both could be compared as both thought of extremely complex networks of treaties and relationship without any successor. In his words, he defined diplomacy as "a king who understand the true implication of diplomacy to conquers the whole world."[52]

Kautilya's concept of diplomacy is important because of the theory of Mandala and its six forms.

Six Forms of Diplomacy

Kautilya suggests that both strong and weak kings to follow certain strategies to protect himself and the state. Diplomatic forms are depended on the king's policy directed towards superior, inferior or equal power. Superiority or inferiority of a king depends upon political, economic, military and geographical powers. The six forms of diplomacy, as mentioned by Kautilya are: sandhi, vigraha, asana yana, samsarya and dvaidhibhava.

Sandhi

It means that king's wanted to accommodate with his adversary and do not want to resort to hostile means. It could be permanent or temporary depending upon the situation and the relative position of the kings. Kautilya mentions five different types of sandhis. Mitrasandhi is based on concrete terms. Hiranyasandhi is related to transfer of wealth. Bhoomisandhi is based on transfer of land or territory. Karmasandhi deals with military exchange. The last anavaistasandhi means to colonize an unoccupied place. These various subforms of sandhi are practiced even in modern time. For example, Bismark had to rely on Karmasandhi with Austria and Britain foreign policy earlier was based on anavasitasandhi with the USA.

Vigraha

Kautilya has given more importance to diplomatic contest than an armed struggle which goes by the name sangrainka.[53] Vigraha, as a diplomatic contest is a mean to achieve the end, i.e., to avoid regular warfare. This demonstrates that even in those times the law of war and peace had gained importance in ancient India. It is interesting to note what Kautilya had said in 'Yatavyavrtti'.[54] Kautilya depicts a picture of diplomatic struggle carried on between attacking and attackable powers and everyone trying to get the best possible result. This diplomatic war ceases by ceding soldiers, money or an ally. Thus, this method was more gainful to the king than fighting a regular war.

Asana

The third form of diplomacy was asana. Some scholars loosely interpret as neutrality. But some aspects of neutrality are implies in the term asana. This issue is quite a debatable one. Henry Wheaton remarks: "according to the laws of war, observed even by the most civilized nations of antiquity, the right of one nation to remain at peace, while other neighbouring nations were engaged in war, was not admitted to exist."[55] According to him a state was either an ally or an enemy. There was no intermediate relations. In ancient India, one finds this intermediate relationship existed. The term madhyama and udasina are used frequently to denote that these terms could be understood as neutral kings. The udasina's territory was the neutral zone where the law of peace prevailed in actual practice.

Kautilya defines madhyama as the one who occupied a territory close to both the conqueror and his immediate enemy in front and who is capable of helping both the kings whether united or disunited or of resisting either of them individually. The udasina is that king 'who is situated beyond the territory of any of the above kings and who is very powerful and capable of helping the enemy, the conqueror and the madhyama king together or individually'.[56] Hence, both the madhyama and udasina occupied interior zones apart from diplomatic zones. A clearcut distinction is not possible between them. But, some scholars believe that madhyama is one a mediator than a pronounced neutral. In certain cases, it is also true that the conqueror entered the madhyama territory but there is no precedent for the invasion of the udasina kingdom.[57]

Yana

The next form of diplomacy is yana. Kautilya uses the term abhyuncayo yanam and Ganapati Sastri believes in Saktidesakala. It means preparedness to march taking into consideration the army, place and time. An invading king could embark on a march after properly safeguarding his own state, if he felt confident that he would be able to destroy the enemy's works.[58] In a later section of Kautilya work, the diplomatic utility of the yana is given.[59] If the conqueror finds his enemy surrounded by troubles and people are not happy and united and his kingdom is faced with pestilence and other kinds of ill, then he may attack vigrahayanam. He could again resort to yan if he found his allies in front and the enemy in the rear of them he must be heavily backed by his own people.

Samsraya

It was the fifth policy under the sadguna policy. This policy of protection is followed where a stronger state intervenes and gives asylum to a weak state. He used this policy when strong needed help to defend itself from equal power. One can say that in the 19th and 20th centuries that in the process of colonialization European powers controlled the weakner nations in Africa and Asia and by doing this they strengthened their position against each other. Samsraya was based on aggressive policy. Kautilya had mentioned peace and stability in a state which made it more powerful and aggressive against weaker enemies. He thought under an unjust king the society was not secured because of social tensions.

Dvaidhibhava

It meant double policy. Dr. Sama Sastri's interpretation "making peace with one and waging war with another" is thus subject to correction. Underneath this policy lies an attitude of duplicity."[60] In the 20 century, Kissinger followed this policy when he made an alliance with China so that at no time the Soviet Union and China could come closer as regards to US friendly gesture to China. Kautilya spoke the samething within the mandala framework.

Thus, it can be stated that Kautilya's diplomacy and foreign policy were derived by the ambitious kings who wanted more and more power and wealth. His diplomatic tactics were equally influenced by Hindu religious and social structure which helped him in the formulation of foreign policy and its application.

Kautilya's method of managing war and diplomacy has admirers in modern age. His sixfold policy of diplomacy is still relevant even though the countries are now separated by oceans and inter-continental ballistic missiles have affected a country's geography. One can say that Kautilyan thinking had great impact on successor writers but in case of India the story was different. The strategies adopted by Kautilya were seldom applied by the Mughal kings and later on by Britain when they conquered India.

Another important question that emerges from this discussion is the applicability of his theory to different forms of governments: democracy and autocracy. Plato, Aristotle, Kautilya and Machiavelli believed in the supremacy of king's rule and the state was an ultimate end. Kautilya believed that the goal of state was peace but this had to be achieved by means and one such means was war. His note on statecraft indicates that a "wise king trained in politics will even if he possesses a small territory, conquer the whole world with the help of the best fitted elements of his sovereignty and will never be defeated."[61]

Instruments of Diplomacy

Kautilya envisaged that the successful diplomacy was made possible by four means: sama, dana, bheda and danda. These are recognised traditional expedients which could also be traced in Vedic writings. It needs further detailed explanation.

Sama: The implications of this diplomatic means of sama are found buried in the texts of the Dharmasastra and Arthasastra.[62] The conquering king is asked to observe this policy of conciliation by the promise of protection of villages and forests, of sheep and cattle, of roads and traffic of the restoration of the banished and of the runway.[63] Kautilya recommended the application of this policy towards the conquered king so that the latter may remain loyal.[64] The purpose of sama was to control weaker kings and if this failed the other methods were adopted.

Dana: Kautilya recommended the policy of sama and dana to inferior kings. The policy of dana was made applicable if direct method of sama failed to achieve the desired goal. It is said in the Matysa Purana that even gods were brought round through gifts of wealth and other presents.

In case of new situations arising when the local chiefs or leaders support the alien forces against the king. Kautilya says that a king should follow the policy of sama and dana towards the discontented subjects so that they might consider aligning with the foreign powers. In this kind of situation, Kautilya says the first act of the king is to suppress the internal danger and then face the external danger.

Bheda: The third expedient was bheda. In Matysa Purana it is stated that bheda or dissension should be pursued towards the unrighteous king. It is the policy of 'divide and rule'. A kingdom can face both the dangers, external and internal, at the same time. Since internal peace is essential the Purana recommends the use of the policy of bheda. Though Manu does not attach much importance to this policy but Kautilya sees it is as an effective policy to bring even a strong king under his subordination. According to Kautilya, by the policy of bheda or danda a strong power can be subdued. Kautilya talks of different methods of sowing seeds of dissension. These are done by instigating anyone of the neighbouring kings, a wild chief, a scion of the enemy's family or an important prince.[65]

Danda: Besides these three expedients, there is an important element which Kautilya calls danda. In Arthasatra, we find a systematic examination of the concept. The word danda is very complex to explain. Primarily it means a staff or rod. The bramachari and the sanyasi were expected to have a danda. In political sense, the king had his danda which symbolizes kingly power and the authority to punish the offenders. In Arthasastra we find danda used in the sense of a staff.[66] The term danda also means punishment. Again it would be interesting to compare Aristotle and Kautilya as they express similar views on punishment. Aristotle believes that "judicial decisions are useless if they have no effect, and society cannot exist without them, neither can it exist without their execution. Therefore, Aristotle accepted that for the proper realization of the state and the progress of the world, punishment was very necessary.

Kautilya recommends physical punishment or fines. Another meaning of danda is army and Kautilya uses the term also in this sense. But the most appropriate meaning of danda[67] is coercion or punishment. Kautilya and even in Mahabharata and other works, the term dandamte or the act of punishment is permissible.

Kautilya's view "who is desirous of the progress of the world shall ever hold the sceptre raised" (udyata danda). Never can there be a better instrument than danda to bring people under control. Kautilya cites the famous law of the fishes and to the general need of protecting the weak against the strong with punishment which have been referred earlier while dealing with state's functions.

While repudiating the view of the earlier writers Kautilya emphasises on severe rather than mild punishment. Danda should be dispensed after due consideration and in proportion to the gravity of the offence. Danda when wrongly inflicted would lead to the creation of evil feelings not only among householders but also among ascetics. If not used at all, it will lead to anarchy. Hence, danda when used after balanced reasoning leads to the growth of wealth, enjoyment and dharma of the subjects.[68] It indicates danda should be used as and when deserved.

The Institution of Ambassador and Espionage

Ambassadors are an important part in the execution and fulfilment of the sixfold policy of the kingdom. Regarding the qualification of the ambassadors, Kautilya has emphasized that they could not be selected from amongst the relatives or favourites of the ruler unless they fulfil necessary ministerial qualifications. They enjoyed the status of ministers. They also enjoyed immunity from arrest. Some care were taken in the selection process such as loyalty to the king, freedom from vices, capacity, honesty, strength of character, eloquence, brilliance, forgiveness and high birth. Great stress was laid on the conduct of an ambassadors at foreign courts. He was expected to behave with dignity and courtesy and to preserve the good name of the same which he represented. When on a mission to an enemy state it was the duty of the emissary to act with courage and resolution and at the same time to show moderation and tact.[69]

Kautilya mentions three kinds of ambassadors. Nisristarthah was the first kind of ambassador. He possessed ministerial qualifications. They belonged to envoy of class A. Parimitarthah was the second kind who possessed the same qualifications. He was for some specific mission were class B envoys. The last category was that of Sasanaharah. He also possessed the same qualifications. Generally, he was conveyor of royal writs. Kautilya made an improvement of Manu's concept, who was silent on this issue.

The duties of the ambassadors were to keep the government fully informed of the activities of the courts to which they were attached. They had good relations with officers of state, were well informed and accustomed to the welfare of the kingdom. In particular, they made it their business to ascertain the number, size and positions of the forts and military stations, the strength of the army and the strong and weak points of the kingdom.

Espionage

Manu had set an example which Kautilya followed to perfection.[70] According to Kautilya, the king was to create spies with the assistance of the council of ministers who had themselves been tried under espionage.[71] From this, it may be inferred that the ministers were finally appointed only after their character and antecedent had been fully ascertained. Further, Kautilya states "assisted by his prime minister and high priest the king shall, by offering temptations, examine the character of ministers appointed in governmental departments of an ordinary nature." A pale reflection of this method still survives in modern governments which ascertain through the local intelligence branches or similar detective wings, attached to the police department, the antecedent and character of persons before the latter are finally admitted into government's services.[72]

Kautilya followed a rigorous system of espionage. It resembles in some way with the totalitarian governments of the modern age. According to him, there were nine kinds of spies. These were:

1. **Kapatika–Chhatra:** He worked under the guise of a fraudulent disciple or one who worked in the guise of Kapitika. Dr. Kane interpreted it as a bold pupil capable of knowing the minds of others.
2. **Udasthita**: He was a sham ascetic who failed to perform the real duties of asceticism. He was intelligent and endowed with pure character.
3. **Grihapatika**: He was a householder who was unable to meet his livelihood but was intelligent and of good character.
4. **Tapasa**: He was one who desired to practice an austere life.
5. **Satrin:** He was a classmate or colleague.
6. **Tiksana:** He was a desperado.
7. **Rasada:** He could administer poison.
8. **Bhiksuki:** She was a mendicant woman.

Spies were also used in foreign services besides helping the king in the conduct of internal affairs. They received salaries from foreign kings with whom they lived as servants. If they helped their own and foreign states in catching thieves, they received salaries from both the states. Spies were set in motion against kings who were inimical or friendly, intermediate or neutral with regard to the 18 departments of the governments.[73]

Diplomacy and Foreign Policy as an Extension of Warfare

As a political realist, Kautilya thought that every state worked to produce maximum power and self-interest. Thus, one finds little implication of moral principles in the activities of states. Kautilya observed that "an ally looked to the security of his own interests in the event of similarity of calamities and in the event of the growth of the enemy power". The decision to go for war or remain at peace was dependent either upon the self-interest or advantages to one's kingdom. The prospects of war and peace were seen mainly from the point of view of profit. Thus, Kautilya says alliances are made not because of good or moral obligations but to promote one's own self-interest as well as the interest of the ally for" "when one had an army, one's ally remained friendly or even the enemy becomes friendly."[74] According to Burton Stein, Kautilya describes foreign policy not in terms of great empire like Maurya, but of small warring states indulged in incessant conflicts such as India had experienced before Mauryan empire.[75]

Kautilya was popularly more known for his Mandala theory of foreign policy. In his analysis, immediate neighbours were considered as enemies and any state on the other side of neighbour was an ally; or the enemy of my enemy was my friend. Imagine a series of states to one's west then start counting with one. States number 1,3,5,7 and so on were probably friends, whereas states 2,4,6,8 and so on would be enemies. Kautilya put this basic principle in different ways but most simply was as one with immediate proximate territory was the natural enemy.[76] He also states that Mandala theory in this way was "with respect to the middle king (he himself), the third and the fifth constituents were friendly elements. The second, fourth and sixth were unfriendly elements.[77]

Kautilya believed in the principle of preparing for war and planning to conquer. Diplomacy was regarded by him as another weapon used in warfare. On the question of friend and foes, he observed, how the king must think and act. The kings are endowed with personal excellence and those of his material constituents, the seal of good policy was the "would be conqueror". Again on the principles of foreign policy, Kautilya regarded neighbouring states as a neighbouring prince possessed of the excellencies of an enemy was the foe; one in calamity was vulnerable; one without support or with weak support was fit to be exterminated; in the reverse case, fit to be harassed or weakened. These were the different types of enemies."[78] When Kautilya wrote of "exterminating" an enemy he meant killing only the nobles. He thought that the best policy towards his subjects and ordinary soldiers was to treat them well and recruit them.

It is difficult to compare Kautilya's Mandala theory with theory of modern balance of power. Infact, he was not offering any kind of balance of power theory in his time. In the 20th century, the practitioners of balance of power relied on it because equally armed nations could deter each other and, therefore, war was deferred. This argument was found, to certain extent, in Kautilya's mandala theory: "in case the gains of two allies of equal strength were equal there should be peace, if unequal war"[79] or "the conqueror should march if superior in strength, otherwise stay quiet."[80] The advocates of balance of power theory believed that the state armed itself so that peace could prevail. On the other hand, Kautilya wanted his king to arm the kingdom inorder to find or create a weakness in the enemy and then conquer. Reading Arthasastra one finds violation of moral principles of war. Like Thucydides, Kautilya regarded the request for negotiation as a sign of weakness, indeed this was a desperate act of weak state who want to survives. A weak king will bargain with a strong one with the hope of a gain equal to his troops when he was in a calamity or was addicted to what was harmful (woman, wine or gambling) or who was in trouble. He with whom the bargain was made should fight if capable of harming him, else he should make the pact.[81]

In actual terms Kautilya regards diplomacy as a subtle act of war and a series of actions taken to make the enemy weak and strengthen own self. The ultimate end would lead to eventual conquest. In Kautilya's foreign policy, in order to work for negotiated peace, a king had still to strike again and again and this should be done in secrecy.

Let us also consider some of the measures adopted by Kautilya during peace time. Even during peace time Kautilya activated his spies and secret agents to exploit the division within a state. Most kingdoms, Kautilya observed, had four kinds of unhappy subjects – the enraged, the frightened, the greedy and the proud. Secret agents can further widen these divisions by inciting these four types of people to act against their king. The opposing king should win over the seducible in the enemy's territories by means of sama and dana and those seducible then the king should resort to bheda and danda.

This was Kautilya's typical way of explaining the term peace. Today, the object of a country is to promote peace and security among nation-states. Even in human nature the element of fighting is predominant over the element of cooperation. Proper cooperation among the countries will lead to the promotion of peace. After the second world war, despite a hostile environment, no country wanted war. For this, these countries established international institutions, which would prevent war and work of peace. Whether this aspect was there during Kautilya's time is quite a debatable issue and needs further re-examination of his theory of diplomacy.

Conclusion

War and the conduct of war are largely influenced by the expansion of science and technology. Since ages we have seen the growing impact of it on society and state. In society and the state, geography, population, government, economic resources are important. There are various ways to protect them. One method, when other methods fails, is war or resort to force or violence. In Kautilya's Arthsastra we find in detailed explanation of military art of war. His military writings are also a bold departure from the past writings and was so important that critics began to compare with modern strategic thinkers. As compared to others, his works were written in a systematic and scientific manner.

We have discussed that war is essential for the protection of the states. It was not only in the case of India but elsewhere in other parts of the world. Even Aristotle, Hobbes, Rousseau, Karl Marx, Machiavelli have regarded force or violence as a means to achieve the desired result. A world without war was even theistically inconceivable so long as one was within the established order. A world without war was not even inconceivable but desirable for those outside the fold, i.e. the Buddhists and the Jains. Emperor Chandragupt Maurya established the Magadha empire as a warrior which he himself was well informed of Kautilya's Arthasastra. Though later on, he ceased to believe in the creed of warriors, he as an exception and later on in Indian history whenever the empire disintegrated new empire was built by the development of military powers. In this sense also, Kautilya's Arthasastra becomes important for those rulers, particular during ancient time, to know the rules of warfare and apply it according to the need and functions of the society and state.

In ancient India, the state was regarded as a concrete institution instead of looking upon it as a necessary or unnecessary evils. Seeing the state in a vulnerable position, the foremost duty of the king was to provide protection to the people. Kautilya believed that: (i) states are always hostile and need to be eliminated with whatever method available; (ii) reasons of state must prevail over all other sentiments; and (iii) there are no limits to absolute sovereignty. Hence, Kautilya accepted the importance of the king and his sovereignty. Kautilya's king was a territorial ruler rather then tribal chief and the main aim of the king was to conquer the whole world. Thus, for this realisation Kautilya relied mainly on his own strength and wealth. His strength was strong military force. He properly understood this aspect of state and propounded many military strategies in the Arthasastra. He does not make much distinction between military strategy and statecraft at several places.

Military strategy is an art of manoeuvring of military forces to support not only the political necessity but also economic and social needs of the state. Hence, the threat and the use of forces become important in realizing the ultimate aim. The essence of strategy is the relationship between the two. But the relationship is quite complex. Sometimes the military factors dominate the other factors of society. Thus, the task of strategy is to overcome the threat factor and establish the foundations so that the state could be secured either from internal or external threats. This kind of protection could be given by the king. Kautilya says that "whenever the king is superior, he shall not waste any time and should proceed against the enemy whenever but doing so the enemy can be weakened or crushed."

One thing is clear from Kautilyan writings that decisive battles are dependent upon favourable conditions and time. But, in modern times, due to the growth of more and more complex weapons, the pace of warfare has increased and new weapons' technologies have also affected the area of combat. Hence, the importance of time and weather conditions and Kautilya's advise for long and short marches still remain relevant today.

On the question of offence and defence as the stronger form of warfare, Kautilya regards this as being dependent upon the circumstances in which the king is placed. For a strong king, Kautilya, like Clausewitz, advices for an offensive action and a weak king should adopt a defensive warfare by developing the forts for the protection of the king and the state. In this sense, one can say that Kautilya emphasizes offensive-defensive warfare. Today, the guerrilla warfare is based on the concept of defensive-offensive warfare. But, did Kautilya actually promote this kind of war is difficult to answer. The guerilla war is a protected war ; while Kautilya suggested that a king should go for short and speedy victory.

Furthermore, Kautilya's observation on the management of war and diplomacy is greatly admired even today. His six diplomatic tools and Mandala concept are still applicable in modern age. Kautilya's views on six diplomatic ingredients were similar to that of Karl von Clausewitz that war was the continuation of politics by other means. Similarly, on war, Karl Marx stressed on diplomatic, economic, psychological aspects and only as a resort to military power. Even today, prior to indulging in war a nation resorts to diplomacy, economic blockade and create psychological impact on the enemy so that war can be won without fighting. One can agree that Kautilya, like modern strategic thinkers on similar lines and military (bala) was the last resort to protect the state and the society. His view of the ally and the enemy, on ambassador and espionage were interesting and certainly not out of date. National interest had always been the principal guidance of all states in conducting their foreign policy. Today, states are separated by oceans and long range missiles and have shirking geographic impact on diplomacy. His work can be certainly be applied during De Gaulle times when the Soviet fear of attacking the West Europe became a serious issue. De Gaulle used similar tactics as proposed by Kautilya. It is a great paradox to see that Kautilya's military thinking had greatly shaped the Western future writing. But in case of India, the strategy adopted by Kautilya was seldom applied when the Mughals invaded from Central Asia and later the Britishers conquered India. The answer to this could be the Mughals and the Britishers came with new strategies and tactics which were unknown to Indian rulers at that time.

Thus, one finds that Kautilyan approach of art of war and diplomacy are still applicable. One needs to understand the society, today, which is fast changing. Kautilya ultimately sought peace but his means were war and believed in the concept "would be conqueror" or "Vijigisu" where his kingdom was at the centre and the most powerful one could not attain peace. Kautilya probably assumed that a peaceful empire could last forever and conflict among smaller kingdoms were common. For Kautilya, the kingdom acts in their political, economic and military self-interest. This was a timeless truth of his science of politics or Arthasastra.

Since diplomacy and foreign policy was just an extension of a state's war, the goal of foreign policy was not the end of wars but rather to ward off defeats and to make sure one was successful in subsequent warfare. For Kautilya, all ambassadors and spies were permitted to adopt any means, whether it was moral or immoral. These means have not changed even today. Thus, Kautilya's view on Kutyudha and Tusniyadha further illustrate the principle of immorality in current international relations. Today, we talk of soft and hard measures adopted by a nation against its opponents. Kautilya, too, speaks in this line when he mentions four expedients of diplomacy – sama, dana, bheda and danda.

Kautilya's analysis of statecraft and war has led to a conclusion that it is a text on strategic culture. A strategic culture comprised of social cultures had to be protected by bala (army) and danda (punishment). Kautilya, like any other realist power thinkers thought of power as means to achieve the ultimate goal. Hence, he says, "a king shall always endeavour to increase his power and elevate happiness". For this a king inferior to another in power shall make peace with him and a superiority will wage war. Thus, power that is the basis of the acquisition of more (force), today, is the main ingredient of strategic culture. Hence from Kautilya's writings one can think in these lines which have assumed much importance in international politics.

REFERENCES

1. M.K.V. Rao, *Studies in Kautilya,* Munshi Ram Manohar Lal, Oriental Publication, Delhi, 1958, pp. 2-3.
2. Kautilya, Arthasastra Book I, Chapter XIII, 23, p. 22: text, pp. 22-23.
3. B.S. Saltore, *Ancient Indian Political Thought and Institutions,* Asia Publishing House, Bombay, Calcutta, New Delhi, Madras, 1963, p. 83.
4. Ibid, p. 83.
5. See G. Bull (trans) N. Machiavelli, *The Prince,* chapter III.
6. See Machiavelli, Discourses, Book I, chapter VI.
7. M.V.K. Rao, op.cit, pp. 83-84.
8. *Arthasastra,* 1.17.51. see R.P. Kangle, Vol. 2, 48.
9. Ibid, BK I, chapter XIX, 39, p. 38: text, p. 39.
10. Ibid.
11. Kautilya, *Arthasastra,* VI. chapter. II, 261, p. 291,
12. F. Gilbert, Machiavelli, Politics and History in Sixteenth-Century. Florence, Princeton, Princeton University Press, 1965, p. 170.
13. L.N. Rangarajan, *The Arthasastra,* Pengium Books, Delhi, 1992, p. 31.
14. See V. Kane, *History of Dharmasastra,* Vol. III, 1946, p.18.
15. *Arthasastra,* Book VIII, chapter 2, BK. 1: Chapter 3).
16. I. Topa, *The Minister as a Kinger Maker,* Kitabistan, 1941, p. 52.
17. Kautilya, *Arthasastra,* Book I, chapter 8. Verse 43.
18. See Kautilya, *Arthasastra,* Part I, chapter 4, p. 83.
19. Ibid, p. 85.
20. Kautilya, Book. II, chapter III, 51, p.50: text, p. 51.
21. Kautilya, Book VII, chapter XI, 297, p. 325; text pp. 321-23.
22. Kautilya, *Arthasastra,* part I, op.cit. pp. 86-87.
23. Kautilya, Book I, chapter XX. 45, p. 43; text, p. 45.
24. See N. Machiavelli, The Prince, Chapter XII
25. B.K. Sarkar, *The Political theories and Institutions of the Hindu,* 1923, p. 215.
26. M.V.K. Rao, op.cit, p. 41.
27. V.R.R. Dikshlitar, *War in Ancient India,* Cosmo Publication, 1999, p. 85.
28. M.V.K. Rao, op.cit, p. 143.
29. Ibid, pp. 144-45.
30. Atharva Veda, Book, X, 5.
31. R.P. Kangle, *The Kautilyan Arthasastra,* Part II, University of Bombay, 1963, p. 150.
32. Ibid, p. 196.
33. Ibid, p. 197.
34. Ibid, p. 198.
35. Ibid, p. 204.
36. Ibid, p. 207.
37. Ibid, pp. 207-208.
38. Ibid, pp. 208.
39. Kautilya Book II, Chapter XVII. P. 436.
40. *Arthasastra,* Book X. 1 and 2.
41. Ibid, 2.
42. *Arthasastra,* Book II. 2.
43. E.J. Passion, *Ancient India,* Cambridge, Cambridge University Press, 1922, p. 5.

44. V.R. Dikshitar, *War in Ancient India*, Cosmos Publication India, 1999, p. 38.
45. Arthasastra, 7.11.34: 358.
46. Ibid, 3.1.41-43, 1995.
47. H. Blochmannand, *Ainis-Akbari of Abdul Fazal*, Calcutta, 14 Asiatic Society, Vol. 1, 1927 Reprint (1993), p. 495.
48. *Arthasastra*, Book VIII, Chapter. 9.
49. Ibid, Book VII, chapter 18.
50. Ibid, Book VI, chapter 2.
51. Ibid, Book, VI, chapter 2.
52. Ibid, Book VIII, chapter 2.
53. Ibid, Book I, chapter 16.
54. Ibid, Book, X.
55. Ibid Book VII. 8.
56. H. Wheaton, *Elements of International Law*, London, 1904, p. 564.
57. R. Shama Sastri, *Kautilya's Arthasastra*, Book VI, chapter II.
58. Arthasastra Book VII, chapter 18.
59. Ibid, VII, 1
60. Ibid, VII, 4.
61. Ibid, 11. p. 238.
62. B. Mukherjee, *Kautilya's Concept of Diplomacy*, New Delhi, Abhinav Publication, 1998, p. 253.
63. V.R.R. Distshitar, op.cit, p. 325.
64. Ibid
65. Ibid
66. *Arthasastra*, Book VII, chapter 16.
67. S. Konow, *Kautilya*, Oslo, 1945, p. 45
68. Arthasastra, 1, 4 in Hindu Political thought, p. 86.
69. Ibid, Book. I. chapter 16.
70. B.S. Saltore, op.cit, 480.
71. Kautilya Book. chapter X-XIII, p. 15-25
72. B.S. Saltore, op.cit, p. 480.
73. Kautilya Book I, chapter XII, pp. 19-22.
74. *Arthasastra*, 8.1.56: 389.
75. B. Stein, *A History of India*, Oxford, Blackwell Publication, 1998, p. 78
76. *Arthasastra*, 6.2.9: 318.
77. Ibid, 7.18.1:380.
78. Ibid, 6.2.13: 318.
79. Ibid, 7.6.3: 338.
80. Ibid, 9.1.1:406.
81. Ibid, 7.7.7: 343.

4

THIRUVALLUVAR

Introduction

The development of warfare is an indication of political, social, economic and cultural growth of a country. War is not only a means to maintain a state's security, unity, consolidation and peace and a tool to execute policies but also a means in increasing the territorial limits of the kingdom for the purpose of enhancement of political institution, pride and prestige. In other words, the history of warfare is closely related to the political, social and economic history as war affects all these areas. In ancient India, there was one such poet whose teachings on military issues had a deep impact not only on South Indian culture but throughout India. The poet was Thiruvalluvar. His work has been compared to Kautilya's Arthasastra.

Thiruvalluvar, the Tamil author of the *Thiruvalluvar* (sacred couplets), was born about 30 BC. The book, which contains numerous poems, is considered equivalent to the Vedas of the Hindu scriptures. The work of Thiruvalluvar, initially known as muppal, is a small book containing 1330 pithy couplets. The book is divided into three parts: (i) On Virtue, 38 chapters, 380 couplets; (ii) On Wealth, 70 chapters, 700 couplets; and (iii) On love, 25 chapters, 250 couplets. The work eulogised as an ethical code of conduct for humanity. The poet has not only dealt with general administration but also codified a clear set of rules to govern people's behaviour in social, political, religious and family circles.[1] Thus, for our purpose Thiruvalluvar's explanation "On Wealth" is important because it deals with the state, king and his methods to govern the state. The major contributions of Thiruvalluvar are not confined to one particular age or country, but, like Kautilya, are relevant for all times and for all mankind.

The question of politics and rules governing warfare are well taken in the second part of the book. The title of the book "On Wealth" was not a new discovery. Kautilya had already written his immortal treatise on statecraft, known as *Arthasastra*. The underlying idea seems to be that wealth cannot be amassed or enjoyed unless there is stable and well-established government. He says "the condition of the rich man is more galling than that of the poor under the rule of the tyrant prince. The vast majority of the rules that are laid down for the guidance of the prince and the minister are meant not only for the prince as the ruler of the state but it is also applicable to all mankind."

'On Wealth', gives a list of instructions to be followed not only by princes, kings and laymen, etc., of ancient times but is equally applicable to politicians, ministers and subjects of present days to form an efficient government. He states that a good state or efficient administration depends on strong army, military institutions and military art and the role of diplomacy in war. State is protected well when there is justice and resources. These provisions have to be made by the king who would protect his subjects from internal and external foes. The king should be skilled in the art of war and peace. Thus, it would be of great interest to discuss in detail about his views on various aspects of war and his contribution to the development of military thinking.

Bases of War

Over the years, wars have been decided by three basic factors: physical, organizational and psychological. The first includes terrain or strategic topography, climate, the conditions of the movements of the armies and strategy for planning the war. The second covers the numerical strength, weapons, battle formations, provisions, trained leadership, etc. This depends on the traditions and culture prevailing among the subject in the concerned state. The third implies the differences in the character and mental development of the belligerents. Earlier, Sun Tzu mentioned five factors and seven elements of war. The five factors are moral influence, weather and season, terrain, command and doctrine.[2] Of these, the second and third are included in physical features, the first in psychological factor and the last two in organizational factors. Thiruvalluvar, too, has mentioned these in his kurals but he explained it in different ways.

Thiruvalluvar has emphasised that a king should have a prosperous army, well terrain and good geography of land, good moral conduct and sound organizational values. The military policy of any state involves political considerations relating to the operations of armies, strategy and tactics. Since war is fought between two sides of organized armies, military policy has to be necessarily enunciated to embrace psychology and organisation. While it is rather difficult to lay down any fixed rules on warfare, it goes without saying that no government can ignore any of these factors. All military plans have to be based on these factors. Accordingly, then, three essential factors underlay the military policy of a state. Military thinkers are well aware of the potentialities of these factors in conducting military operations.

Military Geography

The first factor is military geography. Military geography dictates the strategy of war planning. Infact, a state is largely influenced by its topography. Thiruvalluvar says "the water on the surface of the land and the rains that falls from the skies, as well as the mountains, rivers and forts make up a prosperous country." (Kural 737). He envisages this in order to explain that security and prosperity are the two vital aspects of a good state. He perceives that only when a country is prosperous, other countries would think of waging a war on the prospering state to acquire his wealth.

Thiruvalluvar's statement is true in the present context. It is said that USA had waged war on Iraq for its surplus oil and not for the reason that Iraq possesses weapons of mass destruction. Thus, in one way or the other, either geography of the country or its wealth attracts the pugnacious politicians or the states to go for war. In Kural, the terms "irupunal" indicates the two sources of water: (i) the surface stretches of water; viz. the seas, the lakes, etc. and (ii) the water and subsoil water which is very relevant even today. Mountains and rivers contribute both to security and prosperity. Forts are important only for security purposes.

In the modern age, Clausewitz, too, emphasizes that terrain offers such features like nature of soil, the configuration of the land, forest, slumps of trees or brushwood, the situation and size of hills, the position, rapidity or depth of rivers, rugged paths, nodal points like junction of roads, distances etc. These either help or impede the march of armies. According to him, terrain offers two advantages: it impedes the enemy's approach and enables the placement of troops under cover.[3] Jomini also, as in Kural 737, holds that the configuration of the land may facilitate a formidable national defence, especially in a country, which is mountainous or covered with extensive forests.[4] Thus, terrain will help the king to prepare for base, selects the most suitable zone of operations and devise the principles of the art of war. The Kural (494) also says that "your enemy will not achieve his planned victory over you, if you give battle from a well-protected place of vantage." To Thiruvalluvar, terrain is one of the important factors of war because of the condition of the ground, the determination of the use of infantry or cavalry and the size of the army. He, then, warns the king to take precautionary steps before choosing the place for attack.

Thiruvalluvar had correctly assessed the strength of the people. He says "although they may lack fortresses and armies it will not be easy, to attack and dislodge people in their own home stands" (Kural 494). In all wars, the invading armies have this experience. If the local people have a well-planned strategy the invading armies will find it difficult to defeat and overcome the people on their own nation soil. The fiftieth chapter on the "Choice of the Right Arena" lists the importance of choosing the place.

Thiruvalluvar believes that it would be foolish to start on an operation without deciding a suitable arena for it. He, further, says that if the place has been rightly chosen then a weak force can defeat the strongest army. According to him, a general has to assess the value of the ground in the selection of battlefield. He who cannot use the ground properly fails as a general. The terrain or the configuration of land can increase the power of resistance by offering obstacles to the enemy and supporting the flank and strengthening the front and facilities to cover one's army. A battleground should have natural defenses, i.e. a region near a forest for protecting the flank or proximity to a river bank for independent supply of water for drinking or filling the moat. Strong, dusty, wet and sandy sites were to be avoided. In these cases, the first would damage the horseshoes or hooves, the second would prevent easy visibility of combatants and the last two would hinder the movements of men and animals and reducing the mobility. The perfect site would, then, be an area which is vast, hard and smooth. It should neither be very close nor very far from habitation and possess independent supply of water.

Thus, according to Thiruvalluvar, war can take place when the enemy forces come to attack another state. At the time of crisis, a state should have a strong defensive force to face the opponent. This is true even today.

Military Organisation

A good military organization includes good recruiting systems as well as organized systems of state reserves, proper administrative instructions and combat staff, suitable system of rewards, a proper system of assignment to command and direct the principal operation of war and a good system for the commissariat, hospital and of general administration. Numerical strength is not the only factor in war but equally important is the organizational factors like general's qualities, battle formation, trained leadership, arms and equipment, military administration and control, discipline, logistics, espionage, diplomacy, etc. These factors determine the efficiency of the army as a fighting force.

It is a platitude to say that as army is indispensable for the working of the state organism. That is why Thiruvalluvar regarded it as the first of the six limbs of the state.[5] Both Manu and Kautilya rank the army as the sixth limb of the state[6] and prefer treasury to the army. The Kural 381 explains the constituents of a good state as follows: "armies, citizen, economic resources, ministers, allies and fortresses – all six make for a lion among kings."

Thiruvalluvar, then, mentions the main functions and related economic and political standards that would characterise in proper functioning of the state. The necessity of an army, for offence and defence, was even acknowledged during the ancient times. The tributes paid by the vanquished as well as the seizure of the defeated king's assets were considered the legitimate acquisition of the victor. He had seen the state of affairs in his own kingdom and that is the reason why an army was the first requirement of a good state.[7] Therefore, he says "a well organized and equipped army of brave and victorious soldiers should be the first valuable possession of a king."[8]

Dr Subha Annamalai in his "Thirukkural Sindhanai" has explained the kural in the following way: "as a child with intelligence, beauty and discipline is the wealth of a home; good and intellectual books are the wealth of a library; so the efficient soldiers are the wealth of a government." He continues whether the hierarchy is one leader or the commander and under him there are sub-commanders, divisional commanders, officers, soldiers and so on.

According to him, an army organisation is said to be that which is provided with an effective training, discipline, self-control, orders, full devotion to the work done by the soldiers, following the footsteps of the commander at all times and fighting spirit. In organising an efficient and effective army, the commander plays the role of the God Brahma. The commander should create an army like "a good army knows no defeat and co-desertion or disgrace induced by the enemy, but maintains a degree of traditional valour.[9] V.S. Aiyer commented that "alone deserveth the name of the army which knoweth no defeat which is incapable of being corrupted and which hath a long tradition of valour behind it."

In short, it is quite evident from Thiruvalluvar's thought though valour is not something that can be brought and kept but it is an unexplainable quality of every soldier. It should be self-created and self-motivated. Nobody can inject valour in one's soul. Those soldiers lacking valour, are no good soldiers. Hence Thiruvalluvar says, "what is the good of mobilising a roaring army of rats which could be hissed off by a deadly snake."[10]

Thiruvalluvar writing on military organization is quite similar to Ziauddin Barani's 'Fatawa-i-Jahandari' (1358-59AD).[11] The common themes are as follows:

1. The rank to the soldier should be properly assessed comprising merit or efficiency otherwise disorder and distress will flow. No command should be given to a man not suitable for it, nor should a man of higher ability be entrusted with a lower job.

Thiruvalluvar also says that "persons of no background or kindred are unsuitable for ministerial positions, for, being unattached they will be impervious to shame." [12] To add this, Thiruvalluvar says, "the choice of the right executive should be decided on the way they react to the fourfold tests concerning Virtues, Love, Wealth and Fear of Death"[13] and in the selection process, a person's worth and his failings should be fairly weighed and choice should be made, if the good prevails.[14]

2. Keeping a larger and well-equipped army in reserve is necessary to meet an unforeseen danger.
3. The army should never be left idle. It must be useful and "employed in its own profession", i.e. engaged in collecting the revenue, guarding the frontiers, cutting forests, building forts and hunting.

Regarding hunting Thiruvalluvar differs from Barani. Barani says, "the kings should promote the strength of valour and powers by means of hunting, excursions against tigers..." Thiruvalluvar belonged to that society where animals and birds were regarded with great honour and pride. This is quite clear from Chapter 33, titled "Avoidance of killing".

Military Institutions

In his well-known book "On Prince," Machiavelli had stated, "the ability and prowess of a ruler (prince) may be legitimately judged by the quality and loyalty of the men immediately surrounding him." Similarly, Thiruvalluvar had stated that before entrusting a man with a position of responsibility he should be put through a set of four trials relating to Bharma, Wealth, Love and Life. Only those who succeed in these tests may be considered to be of proven faithfulness to the king and the state. Then, the king entrusted them with position and authority. He says, "the choice of the right executives should be decided on the way they react to the fourfold tests concerning Virtue, Wealth, Love and Fear of Death."[15]

Similarly, traces of Kautilya's *Arthasastra* also help in explaining Thiruvalluvar's couplet. Kautilya says that a person should be tested by the purohit, the commander-in-chief, an old relations of a rich and beautiful maiden and a minister. If the person is right, he is selected and given the responsibilities. Furthermore, in his ten couplets, Thiruvalluvar also thinks in terms of the choice of proper men and assigning them appropriate works. He also mentions of recruiting persons for defence services, though he did not devote any chapter on it.

In choosing men, Thiruvalluvar adds some more qualities that are needed in a person to be chosen. He says, "a man of good family background, free of faults and sensitive to public censure will be the right man to choose."[16] Moreover, a deserving and capable person can be appointed if he is not from a reputed background. Thiruvalluvar makes it clear by stating, "even among men of rare learning and virtue, on close scrutiny some patches of ignorance could be spotted."[17] He has given a set of guidelines to both the appointer and the appointee. These are:

- Do not choose men who have no families or relatives, have no social ties, do not fear social blame, and therefore, cannot be depended upon.[18]
- Do not choose men on the basis of affection or for being mere relatives. In short, avoid nepotism.[19]
- Do not choose a strange person for a highly responsible position.
- This point is also for the appointer to be kept in mind after appointing the person, to not fail to trust the trustworthy.[20]
- Trust the person after the appointment.[21]

It is essential to mention that these Kurals are also related to the recruitment of defence personnel.

Types of Armies

Military organization is related to social and political development of a state. In ancient India, many types of armies have been mentioned. Thiruvalluvar had witnessed six types of armies during his age but has mentioned only one and besides this also refers to 'chaturangapala'. In his period, commanders were called 'Thalaimakkall' and the soldier 'Nilaimakkall'. It is generally assumed that a king having infantry, cavalry, elephants and chariots were great. Thiruvalluvar must have accepted this division of army. He says "well organized and equipped army of brave and victorious soldiers should be the first valuable procession of a king.[22]

Parimel Azhaghar mentions of 'fourfold army' for the word 'uruppu' of this Kural in his explanatory notes. In general, there is no clear evidence of the presence of the fourfold army in Thirukkural. R.P. Sethupillai in his work 'Padaithiran' (Efficiency of an Army) explains that there were six types of armies used by ancient kings. These were:

1. Moolapadai (maula – bala – hereditary troops) were the standing army of the state which the kings utilised for offensive and defensive wars.
2. Koolipadai (Britaka – bala – hired troops) were mercenary armies coming from different regions.
3. Naatupadai (Sreni – bala – troops belonging to corporations) was a kind of army which was forced into service by the king when necessary or as corporation of soldiers of companies of men of the same category under their own commanders. Such a force was not always reliable.
4. Kaatupadai (The Atavi bala), literally, were forest or wild tribes or hill men, e.g. Sabarar, Nishadar, Mlechchas under their own chiefs. It constitutes an irregular force of aboriginals and tribal.
5. Thunaipadai (the Mitra bala or Suhrid bala) were armies of a feudatory or an ally. They were unreliable forces at times.
6. Pagaipadai (Amitra bala or Aribal, Dvisadbala) were captive or enslaved soldiers of the enemy or those reduced by the aggressor state or deserters from the enemy. They were used to weaken the latter but they were unreliable.

Among these, only the hereditary troops have increased in number however, and was considered to be most important troops. Thiruvalluvar envisages the hereditary troops as Tholpadai in one of his couplets. Apart from these kinds, there have been elephantry, cavalry and army fronts. Evidences of these can be traced from the following Kurals. In the chapter 'Military Pride', Thiruvalluvar mentions about the elephantry as "a soldier's javelin" wounded an elephant, which ran away with it but looking around for a weapon, he spotted another javelin stuck in his own body and smilingly retrieved it.[23] Similarly, in the chapter "Wicked Friends" he mentions about the cavalry as" it is better to be alone and not befriended than have fair weather friends alone who are like unbroken horses than fail in battle."[24] There are references about infantry too. He says that "on an appreciation of the disposition of enemy forces a good army is always on the offensive, keeping the enemy on the run."[25]

Primal Azhaghar has used the word 'thar' for army front in his explanatory notes of 'Thirukkural'. The term reflects the theme describing the boast of a warrior to his king about his ability to destroy the enemy's front. In the explanatory notes of Deva Neya Pavanar of this Kural, he mentions in brief that the soldiers were placed in arrays of club, division, wheel, chariot, lotus etc according to the situations. There were five special categories of army in Tamil. Netri, Thar, Kai, Perani and Koozhai. The term Netri refers to the front part of the Thar, i.e. the guards of the main army. Kai refers to the flanks of the main array of soldier. Perani is the main soldiers and Koozhai is the rear of the army.[26] It becomes quite clear that during the time of Thiruvalluvar infantry, cavalry and elephantry constituted the main force. Chariot is also mentioned in the chapter 'On Choosing Right Time'. He says that "the strong wheeled chariot are driven on the high seas so, too, the swift sailing ship cannot be propelled on land."[27]

Principles of War

Unity of Command

This is a fundamental axiom of war. The success in any military operation cannot be envisaged without the active and willing cooperation and coordination among all those forces and factors responsible for its realisation. It is comparable to a human body which need active and willing cooperation of all its limbs, nerves, tissues and other organs to keep it alive and moveable. The cooperation has been recognized since the early period of warfare as an essential factor for sustaining the war effort. Today, the increased inter-dependence of army, navy and airforce has to be cultivated and indoctrinated so as to make it a normal routine.

A historical analysis of the nature and content of war indicate that the absence of cooperation either between the national policy and strategy or between strategy and tactics or the inter services and intra-services have proved to be as expensive for the state as to compel it to accept the terms of peace as dictated by the adversary. The wars preceding the second world war were based primarily on the combined operations, making the war to be more complex so far it contents and nature were concerned and more indecisive in terms of its consequences.

A correct application of this maxim of strategy helps in the smooth functioning of the military machine. The significance of command and unity of purpose are the most effective instrument to develop adequate cooperation at different levels in the conduct of war because "only through effective cooperation that components of any force or nation can develop the full measure of their strength" as "better the cooperation and coordination, the greater the chances of success." To make it more effective it can be said that the effectiveness of the overall national war potential with regard to its combat power would depend on the "complete, effective and complete cooperation between all components of the nation, its allies and its military forces." It is the unifying principle of war aimed at evolving a common endeavour.

Realising the importance of cooperation Thiruvalluvar actually criticises the non-cooperators as "only those who know and practice social cooperation live their life. The others are as good as dead."[28]

Army Officers – Commander-in-chief

The organization and methods of military command differ in different states but everywhere it encompasses all activities and problems connected with the preparation and conduct of war; to consider the aim and nature of the war: to direct the preparation, mobilization and deployment of the army and to organize military operations.[29]

In modern times, a good general staff to advise a General is considered to be the most useful part of the military organization. The essential qualities of a commander in-chief are as follows:

1. Courage, both moral and physical, is the foremost quality of the general. The general must be capable of great resolutions and have no fear of danger. Personal bravery was more valued than strategic skill.
2. A general must be imbued with personal qualities like gallantry, sense of justice, firmness, uprightness, capacity to appreciate and utilize and not be jealous of merits of others.

3. A general must possess great character.
4. A general must be thorough and a perfect master of military strategy.
5. Besides intellectual capacity to know and understand strategy, a general must have the capacity to control the human factor, i.e. the leadership character.

Thus, there must be a combination of intellect and natural leadership. The army is inspired by his deeds. It is rooted by his fall.[30]

Similarly, Thiruvalluvar has given some essential qualities but he has not pointed to one particular quality. He has allocated the requirements to person like minister, spy and even king. The first among these is been applied to the king and his minister. The second and fourth aspects are also to the minister. The third pertains to all professions. And finally he takes the fifth aspect in Kural 517. Hence, the characteristics of a commander according to him are:

1. Courage, generosity, knowledge and perseverance in boundless measure are fourfold traits of kings.[31]
2. Resoluteness, heritage, welfare of the people, study of statecraft and driving energy are the five traits needed in a minister.[32]
3. If a man abandons rectitude and covets another's property, he will fall into cumulative errors and disgrace his family.[33]
4. What is there that is too subtle for men who add knowledge of books into their natural intelligence.[34]
5. Even if one is well versed in the theory of book knowledge, it is important to gather experience in practice before one goes on to act.[35]
6. Having examined and decided that this job could be competently done with these means by this person he may be left free to do it on his own.[36]

Council of War

Thiruvalluvar says that "wisdom is the ultimate and impregnable defence before for protection against destruction. It is also the fortresses of inner strength against the enemy onslaught."[37] Kamandaka advised consultation with wise and loyal men with experience of military science, as the power of ministerial deliberation was superior to brute force. But the war council was nothing but an advisory body.[38] References to Councils of War and Military Consultations by rulers or generals, followed by the issue of instructions before the start of the campaign, would indicate the existence of some sort of practice of strategy and tactics.

Leadership

The role of leadership or generalship cannot be minimized in any war. Once Napoleon said "in war men do not count; one man is all important." Thiruvalluvar also expresses the same in the following words, "even when the army had plenty of brave and good soldiers, but without a good commander he cannot prevail."[39] Furthermore Napoleon said "there are no bad soldiers: only bad generals." Thiruvalluvar also agrees with it. Thus, the rank and file of an army unit may be compared of really good and courageous men but if the unit commander is not a sound man then the regiment cannot really make a mark on the battlefield.

Compulsive compromises eschew such advice but this unprecedented anthology of military wisdom confirms a stark and simple, principle of effective military leadership. In war, 'winning is everything.' Even when they were losing, they have never lost the sight of the principle of 'winning is everything.' They also share one cover riding trait: a desire to "get their boots muddy" on the front. Together, their unrivalled understanding of the art of command and their willingness to lead from the front earned them respect and loyalty of those they led were inspired to risk death. The inspired leaders ranged from Frederick the Great to Norman Schwarzkopf. In recent times we had Moshe Dayan, Mao Tse Tung, Vo Nguyen Giap, Manekshaw, Harold Moore etc. All were effective in their time and in their war with the troops they led. Napoleon said on seizing opportunity; "profit by all opportunities, for fortune is a woman. If you lose her today, don't expect to get her back tomorrow." Similarly, whether cruel or kind, fighting for the right for the wrong, what they have to say is remarkably consistent and proves that in matters of military command and virtually any other kind of leadership – the men and the causes may change, but the principles of victory remain the same.[40]

Psychological Factor

Besides physical features, warfare presupposes certain psychological factors. These aspects are mainly associated with the mental factually of the people. Hence, human nature becomes important in cultivating morale and spirit among the subjects. Even the best military organisation is not devoid of this aspect of warfare. The Roman military system fostered the union of civic virtues and military spirit which won glory. But when they lost they gave up military services to the mercenaries and the fall of the empire became inevitable. The army must be invested with due consideration in social and public life and preference was given to military officer in appointments. Thus, the fate of the king is largely, influenced by the morale of the kings and the senapatis and the soldiers. A benevolent and just ruler wields a superior moral influence on his army. If he win the confidence of the people then he makes them more royal towards him. Kurl 544 states "the entire world will be at the feet of the king, who rules justly and benevolently, in accordance with the fair wishes of his people."

This shows that if the king and his subject have good relations and the society flourishes without fear or danger of any kind. The above Kural is made more comprehensive when this is compared to Rousseau's view that the best way for a king to become "an absolute monarch" is to make himself loved and respected by the people. This gave birth to his theory of social control. In this theory, the monarch or the government is responsible to the people because the selected representatives are there to rule over them. If the king failed in his duty, the people had the right to remove that king and select another monarch or government. Furthermore, Machiavelli has remarked that "the sinews of war are not made of gold but they are the gallantry of the soldiers.[41] It shows moral, in a sense, is more important then physical courage and external sources.

Military spirit should not be taken as enthusiasm of the army. Military spirits springs from military institutions and depends upon general's will. The government must reward and honour the deserving soldiers with honour and respective rank and punish who are coward and disgraceful."[42] These are well expressed by Thiruvalluvar as such: "a well organised army will surely win, if it is not subjected to mean treatment and ill payment, resulting in pettiness, incurable disaffection and penury.[43] Further, he says, "in the protection of his people and justice delivery by punishing the wrong doer, is not blot on the king, it is his bounden duty."[44] In other words, a small army with high morale and the king providing them well will produce great victory and on the other hand, if the soldiers are not treated well it will result in loss.

Furthermore, Thiruvalluvar states that when the king does not reward the deservers and does not punish the wrongdoers then the state would go uncared. Though these couplets are some of the guidelines for the just king, these are also a applicable for enhancing the morale of soldiers. Soldiers must feel that the army would become respectable and win glory only through faithful performance of duty and bravery. They must realize that firmness in defeat is more honourable then jubilation in victory. The individual soldiers must possess moral qualities like courage, audacity and self-sacrifice.[45]

History is full of examples where morale was a significant factor and it could be compensated for numerical strength. The first and the third battles of Panipat amply illustrate the effect of the loss of the morale of the enemy's army and as a result Babur and the British scored victories over his opponents respectively. The pertinent question is how morale of the army can be raised? The answer is either love for the state or from fleeting passions, political or religious. Morale and spiritual enthusiasm may be further strengthened by strong discipline within the army. Religious zeal infuses a new spirit into the morale of armies as the sword was considered to be "the key to Haven and Hell."[46] It indicates that an inspiring ideology could become more an important factor for victory than military training and equipment.

There is another way in arousing the morale among the subjects and the king, i.e. when there is proper environment. A king has to assuage the excited passions of the enemy and to foster the same enthusiasm among his subject when they face the enemy. Here, propaganda becomes important in war. The king has to appeal to the psychology of the men and consider their passions. This would also depend upon on the degree of activity and boldness of the army commanders. Besides, these soldiers with a strong will and diligence that goes with it will not give into a cry of despair, moaning that they will be remained in the battle.[47] Once

Winston Churchill, the then British Prime Minister, during second world war, roused the spirit of the people by his oratory after the Dunkirk disaster. He said "we shall cot flag seas and oceans, we shall fight with growing confidence and growing strength in the air, we shall defend our island whatever the cost may be. We shall fight on the beaches, we shall fight on landing grounds, we shall fight in the fields and in the streets, we shall fight in the hills, but we shall never surrender."

Since one of the aim of war is the maintenance of morale, Liddell Hart, then described it that "not battle but the defeat of the adversary" are important from psychological point of view. Even Clausewitz made a statement by saying that "everything in war is very simple" but "the simplest thing is difficult." This makes all the more important to understand the psychology of the soldier during abnormal situation like, the war. Thiruvalluvar says "the stalk of a lotus blossom grows long enough to project the pretty flower out of water, so too, a man's level of greatness is determined by his own will."[48]

Thiruvalluvar explains this with a simile that the flower lotus grows up to the height of the water in the pond. The Kural can be interpreted that only a man or a soldier does his job with full involvement and enthusiasm he can achieve success and then Clausewitz's statement "simple is difficult" would become simple when a state is face with situation of war.

Strength of Army

The strength of the army is much depended upon military spirit and morale. Besides this psychological factor, a state should have a permanent army of respectful size which are trained and disciplined. But in practice, the size and nature of the armed forces are determined by the political policy and the national economy of the state. Hence, the strength of the army differ from state to state, varied from time to time, according to the needs of the situation.[49]

Though numbers are important so far the strength of the army is concerned, Thiruvalluvar never considered it as an important factor. But some of them are mentioned by him as follow:

(a) If a large army is constrained to operate in an area meant for a small force they will have no place for maneuver and so get demoralized.[50]

(b) What is the good of mobilizing a roaring army of rats which could be hissed off by a deadly snake?[51]

War Finance

Thiruvalluvar in the chapter "Ways of Making Wealth" says," make wealth for there is no sharper sword with which to destroy the pride and confidence of the enemy".[52] This verse is couched by the poet in the form of an injunction to the king who is planning aggressive campaigns. In recent times, one can recall that modern management in the form of M.M.M. i.e. men, money and material which are the basic requirements of any venture. And of these, money is the central one for maintaining and enlisting men and for buying and using materials.

War implies expenditure. It requires serious financial estimates, provision of funds not only for disbursing pays and allowances, rewards to officers and troops but also for supply and maintenance of beasts of burden, for equipment, arms, armour and dress, for transport and food and fodder, for advances, bribes and contingent expenses. Thus, war finance plays a most important part in warfare.

Rewards and Distinctions

Rewards, honours and distinctions for good services constitute a regular feature in military administration in all countries and in all ages. Thiruvalluvar too speaks of this in this form; "a well organized army will surely win if it is not subjected to mean treatment and ill payment resulting in pettiness incurable disaffection and penury."[53]

Assessing the Strength

It is pertinent to know the strength and weaknesses of an enemy and of a King or even a Prime Minister when a kingdom is faced with external threats. The strength and weaknesses of the enemy will depend on information. Thiruvalluvar says, "a king will assess the force he needs, the force at his disposal, the enemy's strength and that of his allies before he launches an operation."[54] In the past Gulf War, 1991, Saddam Hussein of Iraq, did not follow this dictum about assessing his own and his allies strength and had to pay heavy penalty for it. If he had done the assessment properly he would surely have listened to the advise of the Secretary General of UNO on the night of 14 January, 1991, and agreed for the gradual withdrawal from Kuwait.

If a king knows the strength required for the operation he embarks upon in relation to his enemy's strength as well as those of his allies and if he concentrates towards his objective final success will be at his door. In this context, Thiruvalluvar says "for a king who is aware of his range of possibilities and studies related problem in full nothing is impossible if he concentrates on his enterprises." [55] Montague is believed to remarked once that "we undo ourselves by impatience." Thiruvalluvar, further, says those who do not know to live in peace knowing their own limitations but are full of self-conceit will surely come to a swift end.[56]

On the assessment of enemy strength, Thiruvalluvar says, "even in jest, it is not wise to entertain the wish to make an enemy of anyone at all."[57] In other words, it is bad thing to go into hostile relationship with others. Furthermore, he says "a fact worse than madness be falls a prince without allies who creates enemies all round him."[58] He praises the prince who has the ability to convert the enemies into allies, the whole world will depend upon his sagacity."[59] Then, further he gives a situation and himself renders the solution to it by these words, "a prince without allies when faced by two enemies will go all out to enlist one of the two for his own ally."[60]

Thiruvalluvar, with reference to administrator, makes it clear by stating, "one's grief need not be communicated to friends who are not aware of it nor is it proper to expose weaknesses to foes."[61] The right thing to do, under such circumstances is to put on a brave front. When the king overcomes his drawbacks with his strategy his enemy is forced to keep at bay. Therefore, he says, "before a prince who plans and conserves his resources and so arrange effective defence the proud enemy will be truly humbled."[62] As such, they will be destroyed who fail to take timely actions to curb the strength of their enemies.

Fortification

(a) **Importance of fortifications:** According to Thiruvalluvar, "fortresses are essential for the kingdom whose ruler is going on the offensive and also for the defence of the realm from invasions."[63] Fortresses are vital for those who perceive themselves as weak and want to ensure the effective defence of their kingdom. It is equally important for the strong princes who feel that their prestige demands the expansion of their kingdom and want to guard against possibilities of counter attack. The security of the fortresses is, therefore, important for all kings. Further, the term 'aran', for fortress, is used in a comprehensive sense, including all natural and build-up defences which was to be integrated into a total system of defence.

(b) Considering the importance of fortresses, Thiruvalluvar has written ten couplets. Military engineering of the art of fortifications constituted an important aspect of military organization throughout centuries in India as in different parts of the world. Fortifications belong to the military art of defence. Defence in open country is more difficult than those defendable through the engineer's art. Regarded as strategic means these serve dual purposes, i.e. to shield the defender and to delay the aggressor to accept at times even make the aggressor defeat. He explains this by saying, "that is a veritable fortress which enables the garrison to defeat the besiegers even when they exert their utmost to overcome the defence."[64]

Thiruvalluvar agrees with Manu's idea that "a single bowman standing on a rampart can fight against a hundred and a hundred can fight against ten thousand. It is for this reason that fortification has been enjoyed."[65] Thiruvalluvar says, "even at the commencement of battle, a good fortress enables the defender's to kill those who make siege and ultimately win the battle."[66]

Fortification also serve certain other military needs such as to safeguards supply depots, flanks and movements, to command roads and rivers and neighbouring areas, to threaten the enemy's communication and afford a sanctuary, refuge or shelter during need or enemy pressure especially when it is large and well provisioned. At the same time, these also sheltered town and villages. The last part can be compressed with the following couplets. He says "impregnability, abundance of life supporting provisions inside and facilities for making a firm stand in defence are the essentials of a fortress."[67] Further, he says "a good fortress should possess every kind of provisions and arms that will be needed and also a garrison of large and brave soldiers who will rise to occasion."[68]

Classification of Forts

Forts were classified on the basis of physical characteristics. Their prospective potential was heightened by natural advantages or great natural obstacles offered by them. Kautilya mentions four types of forts. There are water fort (audaka durga), mountain (parvata durga), desert (dhavana durga) and forest (vana durga). Further, he recommends the errection of a chain of natural forts or strongholds along frontiers under the warden of the marches. It seems that Thiruvalluvar must have read Kautilya's Arthasastra. Hence, his view is similar to those of Kautilya in describing the details of constructing fort. He says "a proper fortress must have a moats of clean water with an esplanade of land with all round lines of sight and backed by hills and woods with shady trees."[69]

The Kural describes in some detail how a king should choose a suitable site and erect his fort. The significant part is the open space, just on the outside of the moat, wherein the advancing enemy may be easily located and effectively dealt with from strategic places on the rampart of the fort. It shows Thiruvalluvar's grasp of strategic and tactical needs of warfare.

According to Thiruvalluvar, height, width, solidity and impregnability are the four requisites that science demands for forts. He says "the fortress walls should be high, broad and strong as well as difficult to approach and attack by the enemy."[70] In another aspect, the whole idea of fortification is to narrow down the places at which the defence potential may by concentrated for the purpose of defending the entire wide areas within the fort. In the process, intending invaders get naturally disheartened in their motivations. Furthermore, he adds one more point to show how the fort should be. "A good fortress is one which cannot be taken by seize or direct assault or even in consequence of betrayal by traitors." In the end, he says that "all the excellences of the fort come to nothing if the commander is not a man of ability in action."[71]

Diplomacy and War

According to Webster's Third New International Dictionary, diplomacy is the "art and practice of conducting negotiations between nations for the attainment of mutually satisfactory terms". Further, it says adroitness or artfulness in securing advantages without arousing hostility, address or tact in conduct of affairs.

It is an accepted norm that war is to be waged as a last resort. Diplomacy and strategy are to be preferred to war. In the 'Art of War', Sun Tzu said long back that "to subdue the enemy without fighting is the acme of skill: annihilation of the army is not the object of military action: weapons are tools of ill omen" to be used "only where there is no alternative."[72] Ramchandra Dikshitar, in *War in Ancient India* says "...to wage war was the duty and privilege of every true ksatriya in several cases the horror of war made the belligerents think of the consequences and outbreak of hostilities by a well calculated policy which we now term diplomacy."[73] He adds that the four cardinal points of ancient Indian diplomatic system – sama, dana, bheda and danda – were effective instruments in averting many wars. There are statements made by Thiruvalluvar with respect to interstate relations and diplomacy. There are several chapters which show the relationship of states and kings in the names of friends, true friends, enemy, etc.

In ancient India, the normal practice was that the state was governed by sixfold policy of diplomacy. In "Thirukkural there are certain Kurals which mention the various methods of diplomacy. The sixfold policy are: Sandhi, Vigraha, Asana, Yana, Samsraya and Dvaidibhava.

Sandhi (Peace)

This was an open procedure in diplomacy. To seek war would be ruinous to both the parties. To undertake an operation against a superior is like a foot soldier opposing an elephant and to make war against an equal is but the collision of baked mud vessels against each other causing mutual destruction. But the inferior must be attacked and made to submit if the differences cannot be amicably settled. There are also cases where an inferior may go for alliance with a superior by offering himself to help the latter with army, wealth or land that is called Hina Sandhi.[74]

Thiruvalluvar, in the chapter 'Mastering Enemy Characteristics' says "it is wise to strike the weak enemy without hesitation with a strong force. It is imprudent, trying to measure strength."[75] Thiruvalluvar thinks that it is a wise thing for a weak king not to go for an ill-advised strike with a powerful enemy. Against a known weak opponent, the strategy should be to of all out in a systematic aggression without relaxation. According to Parithiyar, critic of Thirukkural, points out going against weak enemy would perhaps also be improper to consider Thiruvalluvar's ethical standards.

Vigraha (War)

Though utilitarian in its nature, Kautilya attaches more value to a diplomatic struggle than to an armed contest. Vigraha is a diplomatic contest but it is a mean to obtain the end, i.e. to avoid regular warfare. It seems that Thiruvalluvar has not mentioned anything on Vigraha or going for war. He talks of ceding soldiers or anybody else. He is against cheating anybody.

Asana

The third method of the sixfold policy is Asana. It is loosely interpreted by some scholars as neutrality. Some aspects of neutrality are implied by this term asana. Henry Wheaton in his "Elements of International Law" remarks that "according to the law of war, even the most civilized nations of antiquity, the right of one nation to remain at peace, while other neighbouring nations were engaged in war, was not admitted to exist." Thiruvalluvar, to certain degree, agree with Henry Wheaton. Wheaton considering the modern trend has said that a country should be neutral while its neighbouring nations were engaged in war. On the other hand, Thiruvalluvar must have taken his society into consideration. He says "in face of misfortune between two enemies whose potential is unknown, it is right not to antagonize either of them.[76]

Yana

The term Yana means preparedness to march with due regard to army, place and time. There are different kinds of Yana. The main are marching with ally (Sandhanaja), attacking from the rear (Parsnirodha), causing dissension between his ally and his enemy (Mitravigrahini), taking the offensive without the crumbling of the defences (Dandvaja), marching with an invincible army (Nirvaja), helped by his enemy's vassals (Kulya) and pretend to march against one's enemy, but suddenly turning towards another (Sighragamini). All these kinds are being mentioned in Sivatattvaratnakara.[77]

Samsraya (Friendship)

As a fifth method of diplomacy, Samsraya, on which Thiruvalluvar has written a number of couplets. His couplets reveal that he partially agrees with the ideals of Samsraya and partly disagrees. He says that "there is nothing so rare to achieve as a good friendship and no better armour against enemy machinations."[78] He has vividly described the peaceful life of a king who has friends of noble quality and friend who corrects him and advise him as when needed. He also criticizes people who become friends only in good times and people who speak out one thing while having in mind just the opposite. He asperses these kinds of men and says that these kinds of friends will bring million time disaster to the king or the administrator. He also held the view of false friendship or deceiving friends.

Isaac Goldberg once said, "diplomacy is saying and doing the "nastiest things in the nicest way." Deception is a part and parcel of diplomacy. From the ancient history to the hi-tech cyber world every state practices diplomacy to achieve their national interests.

Dvaidhivbhav (Duplicity)

It means policy of inducing in two enemies confidence by mere speech but behaving inimically in secret. This policy is to be adopted only when help is not possible from any other source. Thiruvalluvar, while explaining this, has written two chapters on dvaidhivbhav. But it is not clear whether he is for or against this method being used as one of the weapons of diplomacy. Since he has mentioned the consequences of having false friendships he points out this by saying, 'the friendship of an inimically disposed person treats you like an anvil to strike on when a suitable opportunity occurs."[79]

V.V. Iyer explains this Kural in his "Maxims of Thiruvalluvar" as "there are some who do not have any love at heart but only feign intimacy, awaiting a chance to hit when on occasion arises. One has to be vigilant about such characters, even initially, at the time of first contracting friendship".

Furthermore, Thiruvalluvar says that "the association with persons who profess intimacy in the private of one's residence and cast aspersions in public may be restricted effectively."[80]

This Kural resembles the Sanskrit term 'Mityakarana,' one of the classifications of dvaidhivbhav mentioned in Sivatavarathnakara. In all the above said couplets, Thiruvalluvar thinks about friends and no one else. But in Sanskrit it is being mentioned that a servant to a king is deceitful person and its helps as one of the diplomatic factors is needed when help is not possible from any other source.

Thiruvalluvar speaks of all these ideals in terms of bringing disaster to persons who trust their wicked friends. However, Thiruvalluvar's ideas on "Friends", "testing Friendship", long lasting Friendship", "Wicked Friend" and "False Friends" are good examples for a kingdom to have these as the principles of international relations.

Kishan S. Rana, in his book *Inside Diplomacy* has brought out some positive features in Thirukkural. He says, "management experts have found that Thiruvalluvar's writing in moral values give tips on good management, delegation of authority and organizational practices. One portion deals with interpersonal communication so central to diplomatic functioning. It is worthwhile to go back to these roots of ancient heritage to rediscover the range of timeless ideas as well as the foundation of one's own culture." He adds that "these texts should be made part of the training agenda for new entrants into the foreign service and for all Indian officials."

Espionage in War

Espionage is an indispensable adjunct of the conduct of war. Long ago

Sun Tzu urged 'know the enemy and know yourself" in a hundred battle you will never in peril."[81] Indeed, well-organized military espionage or strategic intelligence in peace and war alike make rational war preparations possible. It supplies various kinds of information about the potential enemy plan's and capacity, ensures adoption of precautionary measures, political, economic and military before aggression and helps in conducting operations.

In ancient India the necessity for getting and using strategic intelligence was fully understood. Spies played an important role in civil and military affairs. They were useful in reconnoitring troops, enemy camps in ascertaining the exact position of the enemy, its forts etc and reporting the latest position. Thomas Hobbes describes intelligence service as "the eyes of the king" just like Thiruvalluvar, On 'Intelligence Service,' says "the intelligence service and the literature of statecraft and laws are clearly the two eyes of the king."[82]

In the Kural the two figurative eyes perform extensive and intensive functions. The Intelligence Service covers extensively the areas and population of the state which the king himself cannot adequately cover through the limited range of his own eyes and ears. On the other hand, the literature of statecraft and law provide him of an intensive insight of situation and problems, through his own inner eye and mind judgement.[83]

Thiruvalluvar distinguishes some important functions of a spy in Kurals 584, 585, 586 and 587. Kural 584 says that it is the duty of the king's spies to watch over even the executives of the king and not merely those who have developed grievances of enmity towards the king. In short, "All the king's men, his kith and kin and those inimical to him should all be covered by a network of effective intelligence."

The Kura 585 and others reflects the essence of some of the functions mentioned by Kautilya. These are:

(i) to help in sowing seeds of dissension and creating propaganda in favour of one's own kind and bribing subjects etc;
(ii) to lure the enemy by various measures;
(iii) to delude the sentinels at gates of forts;
(iv) to seduce wild tribes with prospects of plunder and power;

(v) to reside as traders and house holders spies and help to destroy enemy's sources of supply, stores and granaries.
(vi) to encourage their own army and frighten enemy ranks.

Thus, Thiruvalluvar's view on espionage being similar to Kautilya's, agrees to Kautilya's descriptions of the functions of a spy and also mentions the important qualifications of a good spy or modern intelligent men. According to him, a spy should not be easily identifiable by appearance. He should be an expert at disguises. When under disguise and otherwise, it should be possible for him to merge easily with the surroundings and ward off suspicion. In any case, he should not get unnerved by a scrutinizing starer. He should be able to protect a secret under all circumstances and never give it away. It is said for a good spy that even his normal appearance should not be distinctive and should never arouse any kind of suspicion.[84]

Thiruvalluvar mentions that "spies disguised as sanyasis and holy men may easily secure access and information from unlike sources and protect, guard their vital secrets even under extreme coercion."[85] Similarly, he says that "good spies develop inside sources for hidden information which they should verify and present fully confirm as unconfused intelligence."[86]

The king was advised by Thiruvalluvar to constantly ascertain the strength of the enemy as his own strength through spies, through display of energy and through actual conduct of operations. He speaks about the art of government in these words "his enemy should not know his weakness but he must know the weak points of the enemy: he should hide the departments and he guards his own weak points."

Regarding certain functions associated to spies Thiruvalluvar says that "it is the duty of the king to get to know immediately everything that happens, to every body in his kingdom at all times."[87] He also mentions about external intelligence. He says "a prince who does not secure effective intelligence through his network of spies is not likely to be successful at all in warfare."[88] Thus, for defensive and offensive operations, the strategy will be shaped by the intelligence envisaged and gathered in advance by trustworthy agents. But the tactics, to be adopted in respect of each operation, will need the immediate intelligence which will be secured by scouts in forward areas.

Only the king who is capable and resourceful enough to obtain both types of intelligence properly and in time could win victory. In addition, the king should not rely on one spy only. According to Thiruvalluvar, the intelligence procured by one agent should not be directly accepted but invariably be verified by the information secured by another agent, unknown to the former, before even considering acting on that. This shows his balanced pragmatism in dealing with statecraft when he discusses these issues.

In this context, Thiruvalluvar says "the information secured by one spy should be counter checked with intelligence procured independently by another spy."[89] Further he says, "spies where more than one is utilised should not know one another where three agents corroborate another, the information is very like true."[90]

Finally Thiruvalluvar ends his chapter 'On Intelligence Service', with these words, "open recognition of the good work done by the spy is out of order as that will expose the man and his methods.[91] The identity as well as his methods of spies should be effectively kept secret and any recognition of their good performance must be done discreetly without divulging their identity or trade secrets. This is necessary because the protection of the identity and methods of good spies are important for the continued usefulness.

Thiruvalluvar and Kautilya

Kautilya and Thiruvalluvar may be called contemporaries. The former was advisor to King Chandragupta Maurya and Thiruvalluvar was an ordinary weaver. Both lived thousands of miles away from each other and were from different castes, social-political backgrounds and cultures. But it is amazing to find that Arthasastra and Thirukkural have much in common. Sometimes, it seems that they have written the books together. Their thoughts centered around the life, role and qualities of a king, upright kingdoms, faithful citizens, the functions of governments, duties of ministers, ambassadors and spies, wealth and a good army and the role of soldiers.

Life, qualities and duties of a king are found in Kautilya's Books I and VI which are on 'discipline' and "the source of sovereign states." According to Kautilya "the king who is well-educated and disciplined in sciences, devoted to good government of his subjects and bent on doing good to all people will enjoy the earth unopposed." Further, he says, "a saintly king shakes of the aggregate of the six enemies like lust, anger, greed, vanity, haughtiness and over joy; acquires wisdom, restrains from the organs of senses; he is disciplined, maintains his subjects; employs goods ministers for assistance and keep away from his unrighteous transactions." If the king is energetic his subject will be equally energetic. In the happiness of his subjects lies his happiness, in their welfare his welfare.... By his good activities, he can achieve his desired ends and abundance of wealth".

Thiruvalluvar describes a king's life, qualities and duties in eight chapters (80 couplets). It was a Tamil tradition to consider the king as the life and soul of society. A king is one who possesses six things: an army, people, wealth, minister friends and a fortress. The king never fails in four things – fearlessness, liberty, wisdom and energy. He is free from pride, anger and lust. He does not praise himself, has self-control, acquires wealth, guard it and distribute to the subjects. He should be accessible and kind to all, caring and protecting the subjects who come to him. His friends should be men of virtues and knowledge. He even considers his ministers as his own eyes. On the other hand, unkind kings are a burden to the earth.

According to Thiruvalluvar, there are several sources of income of the king. These are unclaimed wealth, taxes which subjects pay and custom collection from foreigners. In his words, there were three channels of equitable distribution of wealth, Defence, Public Works and Social Service. All the qualities and duties primarily attributed to the king in the Arthasastra and the Kural are equally applicable to all people.

On employment of ministers, Kautilya says "sovereignty (rajatva) is possible only with assistance. A single wheel can never move." Therefore, a king shall employ ministers and councillors and listen to their advice. The size of mantra parishad are varied. Men whose ability is inferred from their capacity shown in work, should be, having taken into consideration the place and time where and when they have to work, appointed as ministers. The king shall examine the character of ministers.

Thiruvalluvar describes in a very similar way like Kautilya. Let a minister be chosen after he has been tried by mean of (i) his virtue (aram) (ii) his love for money (porul), his love for sex (inbam) and (iv) his fear of losing his life (uyir). He says a minister is one who makes excellent consideration of time, means, place, manner and the difficulties. He excels in firmness, knowledge, perseverance and protection of subjects. Ministers should possess (i) power of speech; (ii) firm in action and (iii) proper consultation.

On kingdom, Arthasastra holds the seven elements of a kingdom are: the king, the minister, the janpada, fort, kosa, army and mitra. The end of the king and his kingdom is happiness which is secured by his strength. The three kinds of strength are power of deliberation, the possession of a prosperous treasury and a strong army and material wealth. For peace and prosperity in the kingdom, the sixfold policy must be applied. Whatever produces favourable results is policy. The king who is of good character is the foundation of policy. In a kingdom, the forts, the finance and the army depend upon the people likewise buildings, trade, agriculture, cattle rearing, stability, power and abundance of wealth are dependent on people.

According to Thiruvalluvar, for the prosperity of the kingdom three factors are indispensable: farmers (land), merchants (capital) and virtues people (labour). His economics was similar to Adam Smith (1776) and Alfred Marshal (1880) who proposed factors of production in the form of land, labour and organization. Thiruvalluvar believes a prosperous state is one in which there is plentiful harvest, industrial productivity with agricultural inputs and consequent abundance of production and wealth.

On forts, Kautilya advocated that on all four quarters of the boundaries of the kingdom, defensive fortifications against enemies shall be constructed; a water fort, a mountain fort, a desert fort and a forest fort. Chapter IV of Book II describes the demarcation of the ground inside the fort and the building within the fort shall contain twelve gates, provided with both land and water and a secret passage. According to Thiruvalluvar, a fort is one which is an object of importance for those who live in it, which has water, plants, mountains, cool forests etc. It has extensive space within which cannot be captured, by assaulting, blocking or undermining. It has excellent heroes to defend and they should be able to overcome the enemy.

On wealth Kautilya holds that wealth alone is important in as much as charity and desire depend upon wealth for their realization. Arthasastra mentions that wealth which, when obtained, increases the enemy's prosperity or which causes loss of men and money, is dangerous wealth; wealth which causes fear from one's people is provocative wealth, etc. Chapter 76 of Thirukkural is a treatise "On Wealth". Wealth acquired with proper means will yield virtues and happiness. Wealth can also be used to build the army to a strong one so that it may not need much more money to be spent on it.

On good army, Kautilya says that the army is one of the seven limbs of the state. The king should himself supervise the army. He says ksatriya caste is the best for the army. His ministers, and priests, should encourage the army. Without army the king is nothing. Similarly, Thiruvalluvar has emphasised the importance of people and army. He says valour and trustworthiness constitute the safeguard of an army. The army that conquers without fear is the chief wealth of the king. A good army is one that stands firm and does not desert. An army cannot survive without a brave general. He says what "if a host of rats roar like the sea? They will perish at the mere breadth of the cobra." Centuries later, Machiavelli in his 'Prince' said "the chief foundations of all states are good laws and good armies.... Where there are good armies there must be good laws."

On Characteristics of enemy, like Kautilya, Thiruvalluvar lays down pragmatic wisdom on ground of expediency when he advises the king not to go in for an ill-advised strife with a powerful enemy. Kautilya advocates under certain circumstances a king can move straightaway against the stronger enemy first, so that a subsequent engagement against the assailable enemy may not be necessary. Kural 893 is in line with an injunction given by Kautilya in Book VII, Chapter III the king do not foolishly go to war against a superior...you will be reduced to nothing as a foot soldiers opposing an elephant.

It is remarkable that quintessence of some of the modern ideas of planning, management and behaviourable science are found elegantly and well preserved in some of the Kurals and some sections of the Arthasastra. The state those days, was primarily the king and so kingship is dealt in great detail. While most of these precepts would primarily apply to king's handling of projects, finance and war, many of the instructions on normal life's requirements will apply equally to educated and virtuous people of today.

Conclusion

The above critical analysis of Thiruvalluvar's military thinking indicates that the concept of state, kingship, army, diplomacy and war were no longer confined to the northern part of ancient India but had profoundly shaped the nature of functioning of state and king in terms of art of warfare in South India. It is true that there has been radical change in the socio-economic, political and military systems. This is precisely because of the advances of science and technology, which has made war more complex and immoral. Infact, in ancient time, all these thinkers gave importance to righteous war but today morality has lost its relevance in the conduct of war and maintaining interstate relations.

Besides this, there is a common theme in ancient and modern world so far art of war is concerned. Every state prefer stability and if there is an environment of instability and insecurity then war or use of force becomes an essential tool. This was applicable in ancient India and is also true even today. So far stability of a state is concerned, if the modern governments follow Thiruvalluvar philosophy and create the objectives of an ordered state relations regarding the assurance of taking security measures against violence or use of force, definitely the relationship among states would improve. The best example is Europe. After the second world war, Europe was a disturbed continent because of military rivalry between the USA and USSR. Once the Soviet Union disintegrated, we saw Europe emerging as one powerful entity focussed on economic security. This could be correlated to Thiruvalluvar ideas "On Wealth" which he regards as an important factor for the state. To sum up, one can agree with his Kural 734 where he says "that is a great country which is free from famine and epidemics as well as harassment by one's foes." The practical wisdom and knowledge abounding in such ancient texts need to be re-examined, re-interpreted and applied appropriately according to current global development. Thus, it would pave the way for achieving a just and peaceful world.

REFERENCES

1. K.R.R. Sastry, Hinduism and International Law, Reveal des Course, 1962(1), Vol. 117, p. 530.
2. Sun Tzu, *The Art of War*, translated by S.S. Griffith, pp. 63-65.
3. K.von. Clausewitz, On War, Section 11, tactics, p. 4.
4. See H. Joimini, *Summary of the Art of War*, ed., by J.D. Hittle.
5. S.M. Diaz (ed), Thirunkkural, Ramanadha Adigalar Foundation, Coimbatore, 2000, 2001.
6. Arthasastra, VI. Ch. I, King, minister, country, fortress, treasury, army and allies.
7. Kural, 381, in Porutpal, on Wealth.
8. Kural., 761.
9. Kural, 764.
10. Kural, 763.
11. Sources from J.N. Sarkar, *The Art of War in Medieval India*, Manoharlal Publishers New Delhi, 1984, p. 31.
12. Kural, 507.
13. Kural, 501.
14. Kural, 504.
15. Kural, 501.
16. Kural, 502.
17. Kural, 503.
18. Kural, 506.
19. Kural, 507.
20. Kural, 510.
21. Kural, 509.
22. Kural, 761.
23. Kural, 774.
24. Kural, 814.
25. Kural, 767.
26. T. Selvi, Thirukkural Kaatum Samudhayasm, Kudumban, Arasu, Jeyaram Publication, 1968; p. 131.
27. Kural, 496.
28. Kural, 214.
29. J.N. Sarkar, op.cit, p.63.
30. Ibid.
31. Kural, 382.
32. Kural, 63.
33. Kural, 171.
34. Kural, 636.
35. Kural, 637.
36. Kural, 517.
37. Kural, 421.
38. Kural, 770.
39. J.N. Sarkar, op.cit.
40. O. Conneily, On War and Leadership, www.thbookserives.com.
41. See N. Machiavelli, cited in P. Paret, Makers of Modern Strategy, Clarendon Press, Oxford, 1986, pp. 11-31.
42. Kural, 549.

43. Kural, 769.
44. Kural, 549.
45. Kural, 766.
46. J.N. Sarkar, op.cit, p. 25.
47. Kural, 593.
48. Kural, 595.
49. J.N. Sarkar, op.cit, p. 82.
50. Kural, 498.
51. Kural, 763.
52. Kural, 759.
53. Kural, 769.
54. Kural, 471.
55. Kural, 472.
56. Kural, 474.
57. Kural, 871.
58. Kural, 873.
59. Kural, 874.
60. Kural, 875.
61. Kural, 877.
62. Kural, 878.
63. Kural, 741.
64. Kural, 748.
65. J.N. Sarkar, op.cit, p. 144.
66. Kural, 749.
67. Kural, 745.
68. Kural, 746.
69. Kural, 742.
70. Kural, 743.
71. Kural, 750.
72. S.B. Griffith, op.cit, p. 77.
73. V.R.R. Dikshitar, *War in Ancient India*, Motilal Banarsidass, New Delhi, 1948, p. 301.
74. Ibid, p. 316.
75. Kural, 861.
76. Kural, 876.
77. V.R.R. Dikshitar, op.cit, p. 322.
78. Kural, 781.
79. Kural, 821.
80. Kural, 820.
81. S.B. Griffith, op.cit, p. 84.
82. Kural, 581.
83. S.M. Diaz, op.cit, 620.
84. Kural, 586.
85. Kural, 586.
86. Kural, 587.
87. Kural, 582.
88. Kural, 583.
89. Kural, 588.
90. Kural, 589.
91. Kural, 590.

5

KALIDAS

Introduction

The importance of a culture lies not only in its moral and spiritual power but also in protecting from external and internal threats. By doing so, the state is free to pursue its moral, social and economic functions. In ancient India, the concept of 'ideal state' was supreme. This has also been discussed by Kalidas' predecessors. The thrust on stable social order with prospect of diversity in unity was made possible by good political and military organization.

Kalidas planned the architecture of the epic that he was about to write. Kalidas had acquired a reputation as one of the literary masters by the 7th century AD. When exactly he began to write is quite debatable but 4th and 5th century seems to be an acceptable starting date for the composition of his works. The king's life, his virtue and vices, the war they had fought and the territories that they had conquered have been well examined by Kalidas in the light of classical kavyas. Kavyas are reflected in the form of poems, dramas or proses and contribute a valuable knowledge of social, political and military events. Hence, a historical analysis of Kavya has to take the ascetic and representational dimension into account.[1] The spiritual and narrative elements of Kavya was one way of gathering social, political and military ideas related to the structure of state, kingship and war. Kalidas's Raghuvamsham is one such great work where the genealogy and achievements of solar kings had been highlighted. Besides, he wrote Kumarasambhavam, Abhigyanamshakuntalam, Vikramorvasshiyam, Malavika-Agnimistram, Meghadootam and Ritusamharan.

Kalidas's writings had Kautilyan influence whether it was polity, moral traditions and customs, protection of the state and military system. Similarly, one can agree that Smritis, which were the manuals of conduct for the individual and society, formed an essential ingredient in understanding the society as such. It shows that Kalidas followed the administrative standard prescribed in Arthasastra and Smritis literatures. He observed the precepts of rajanitisastra. Therefore, Raghuvamsham becomes important in assessing the political relevance of society, kingships, war, conquest, diplomacy and foreign policy.

Consolidation of the State

Kalidas deals with the three royal powers as spelt out by Kautilya. Kautilya in consolidating the state, described three powers of a king: resources, diplomacy and intrigue and enthusiasm and organization. These different powers are important for the welfare of the subjects and its kingdom. With these three powers, the king and the subjects held status in high esteem. With the accumulation of such powers, the king expanded his empire. The proper utilization of mantras, utsaha and prabhusakti; as mentioned by Kautilya, led to success in every field. In Kalidas's Raghuvamsham, the famous Raghu is bestowed with these powers.[2] Accompanied by his three queens King Dasartha resembles God Indra who is embellished with these powers.[3] The three queens have been compared to three powers. King Aja took full advantage of his prabhushakti.[4] With the timely use of this power, he was able to bring the enemy kings to his jurisdiction. King Atithi is also regarded as the embodiment of these three powers.[5] The proper use of these powers enabled all these kings, from time to time, to expand the territories without humiliating the subjects. This indicates the consolidation and the rule of a state by a king who was more or less respected by the subjects.

Kingship

In ancient India, the principle of divine right of kingship was quite popular in ancient India. The majority of Smritis and Puranas have focused on the functional resemblance between the king and the deities. It meant that the king himself is not god incarnate but his function is similar to that of gods. Kalidas, too subscribes this view. He described Dasartha as a king who by his even handedness, by the showers of riches, and by keeping the wicked under control was equal to Yama, Kuvera and Varuna respectively. His splendor resembled the sun.[6]

Hindu kingship was based on prosperity, security and growth. Under an able king, the kingdom flourished and this was major theme of Raghuvamsham. A dutiful king had to work in accordance with the subject's will and not his own interest. Kalidas refers to Rama as a king who went to the extent by sacrificing his self-interest and personal comforts to make the subjects happy and prosperous. Kalidas also emphasized on the principle of public opinion. It is evident from the fact that the banishment of Sita by Rama was in accordance with what the public felt when she was made captive by Ravana.

Kalidas's account of kings qualities is similar to those qualities which are mentioned in Arthasastra and Dharmasastra. In Kalidas Raghuvamsham, the epithets, attributes and action of the Iksvaku kings became a powerful embodiment and expression of the ideas of mid-first millennium Indian kingships.[7] Martiality, rituality and benevolence were the essential attributes of kings. Kings Dilipa, Raghu, Rama and Atithi were endowed with these qualities. Out of them, Kalidas regards Raghu as the real hero because he corresponds most closely and directly to the ideals laid down at the beginning of the Kavya. Rama is also important, but unlike Raghu, he is god king who can be partially emulated by his human counterparts.[8]

The important elements in Kalidas's construct of ideal kingship are laid forth in five verses where the traits of the members of Raghuvamsham are enumerated. Kalidas has enumerated the quality of purity which all the solar kings inherited from birth. This shows that the aristocratic values were important for the king and his position was such that the subjects bestowed respect and honor. A king should be valorous but should not harm any innocent people. The prime duty of the king is the protection of the kingdom so any kind of fear does not bother the king. Hence, valour and statesmanship were the greatest assets of the king and for such a king the army, though kept ready, was hardly requisitioned to be on active service.[9] Kalidas mentions about the powers of the king and compares the powers of Raghu and Dasartha with the brilliance of the sun and god respectively.

Another attribute of a good king is that who controls his senses. According to Kautilya, this is motivated by training in the science and should be free from lust, anger, greed, pride and arrogance. The whole science prescribes control over the senses.[10] Kalidas has given sufficient importance to it and recalls that the kings possess knowledge of science. By this power, the king was able to properly assess the nature of politics. Moreover, the knowledge of sastras had helped in controlling the senses on many occasions. Kalidas refers to Kings Dilip and Dasartha who were not entrapped by the objects of the senses. Kalidas believed that the internal enemies marked by lust, greed and anger are more dangerous than external enemies and, therefore, must be dealt firmly by the king.

According to Hindu ideas, the end of worldly existence is dharma, artha, kama and wisdom. Dharma is defined as duty, an obligation to be discharged for the good of mankind. Artha is the material aspect of life, object, purpose and wealth; the honourable and honest means to acquire the same. Kama is desire, greed and lust. Without kama there can neither be motivation nor creation. Wisdom and good conduct lead to moksha, liberation from the cycle of birth and death. Kalidas was well aware of these four streams of knowledge He describe Atithi, the grandson of Rama, as maintaining a sense of equipoise in his attitude towards the realization of these objectives. None of these is ever in conflict in his undertaking.[11]

But, in case of contradiction and conflict among them, the choice is left to the person. In ancient India, there prevailed different schools of thought to solve the difficult problems. Dharmasastra and Smritis have given importance to dharma. Manu, the lawgiver, prefers dharma over artha and kama.[12] Vyasa in Mahabharat also attaches importance to dharma because it is the source of both artha and kama. In Santipurana dharma is regarded as the best, artha is the middling and Kama is the lowest goal of life and a man should follow dharma as the principal aim of his life.[13] Even Bhagavata Purana and Yagnavalkya also agree with them.

In Arthasastra, Kautilya lays emphasis on Artha. Artha should be the main goal of human life. Dharma and kama are rooted in it.[14] To Kalidas, dharma is more important than artha and kama. It seems he was greatly influenced by Ramayana. Hence, representing the dharmasastra ideals, Kalidas characterizes King Dilipa as a king who gained wealth and gratified his desires thinking them as part of dharma.[15]

Furthermore, Kalidas says that a king should be trustful. He considers it an important trait and accordingly has mentioned this not less than four times in Raghuvamsham. Kalidas felt that for the sake of truth, king should not indulge in too much talk.[16] He should not speak untruth and whatever statement a king should make must not be separated from truth.[17] Kalidas says that being a righteous king, he should not avoid unrighteous people. A rightful and dutiful king is much respected by his subjects. Kalidas's Dilipa was a righteous king because he followed the policy of righteousness. He regarded a righteous enemy as his friend and a wicked friend was discarded. Besides, this righteous impartiality was another quality of a good king. King Aja was well known for this quality. Charity, too, was regarded as one of the essential qualities of a good king. Kalidas mentions that Raghu, after conquering all the four quarters, performed the Visvajit sacrifice and gave away all his wealth as dakshina.

Duty of a Ksatriya King

The foremost duty of a ksatriya king was the protection of his subject. Kautilya means one who protects people from destruction.[18] Kalidas mentions the sacred duty of a ksatriya king as follows: "there was no fear of theft in the kingdom of King Dilipa since theft having turned away from other men's property, remained merely as a term to be heard off.[19] In the absence of theft and similar other offences in his kingdom, Dilipa had no prisoner to release in the happy occasion of the birth of his son Raghu.[20]

Similarly, Kalidas highlights that in the reign of king Dasartha no disease has set its foot in the land. In other words, considerable measures were taken in the area of health and sanitation. King Rama was also free from greed and avarice and his subjects were contented and did rightful things. These are indications that the lives of the people were well preserved and as such the kingdom flourished.

Relationship Between the King and Subject

Kalidas was of firm conviction that there was an ideal situation between the king and his subjects. Under an able king, the subjects prospered and the kingdom was secured from any kind of threats. The inherent good qualities of a king inspired the people to serve the kingdom in a meaningful manner. In Raghuvamsham, King Dilipa is depicted as the virtuous father of his subject.[21] He regarded royal glory as his mistress.[22] He was always alert in protecting his subjects. Similarly, Raghu pleases his subjects by his benign rule. In real sense, he justifies his title as "rajan"[23] by making his subjects happy. Dasartha was a noble and worthy king. As god Indra cherishes the farmers with timely rains, so he helps the worthy people by bestowing wealth on them. He viewed all people alike like god Yama, gives charity in plenty to the needy like God Kubera, punishes the offenders in proportion to their crimes like God Varuna, and bears the irresistible lustre like god Sun.[24]

Kalidas's supreme ideal of a king is embodied in Rama. Rama was regarded as a people's king. Even as a boy, Rama killed the formidable demons in order to save the sages. The entire period of his exile was spent on serving the sages and people from the evil forces. When he became king, he looked after his subjects carefully and served them with great devotion and faithfulness. For the happiness of his people, he even banished his pregnant wife. Kalidas says that Rama's subjects had a father in him on account of his being their guide.[25] Similarly, King Atithi was an able and kind ruler. Under him, the people were well protected and made rapid stride in all spheres. His aim was to make the people content. He never neglected dharma for the sake of wealth. But at the same time he did not ignore wealth and desire. In this way, Kalidas follows the Kautilya tradition. Kautilya ordains that subjects when impoverished become greedy, when greedy they became disaffected and when disaffected they either go to the enemy or themselves kill the master. Therefore, a king should not encourage greed and disaffection among the subjects which may lead to his downfall. In the pursuit of dharma, dharma and artha are not an obstacle. In other words, the king should always keep a watchful eye over his subjects as a father does over his son. Thus, Kalidas presents the kings as the father of the people. Mutual cooperation and support between them will lead to stability, prosperity and security of the kingdom.

Minister

Minister is regarded as one of the seven elements of state. In ancient times, kings had a high regard for their ministers and these ministers generally proved loyal to them. They were like friends of the kings. Kalidas mentions Saciva in Canta IV.[26] King Aja associated himself with ministers who were proficient in politics.[27] According to Kautilya, the king should rule over his subjects assisted by his ministers, the deliberative council and eighteen departments of the state. Being the central head yet he consulted his ministers and the people's representative over the affairs of the state. The king derives his power from the will of the people and, therefore, he has to keep them contented. Moreover, for the peaceful and disciplined administration the cooperation of the ministers was necessary. The administration is like chariot. The king and his ministers are the wheels. As a chariot cannot function with a single wheel, the administration cannot run smoothly.[28]

The ministers occupied an important place in the administration, constitution, executive and legislative bodies. The ministers had full power to choose successor to king in his absence. They have the right to dethrone any unworthy king. Even Manu mentions that ministers used to play a significant role in the administration of the kingdom. When the king felt uneasiness and got tired in public affairs, he appointed a minister belonging to a high refuted family to look after the affairs of the state. Yagnavalkya also asserted this truth and mentioned that the king ruled his kingdom with the support of his ministers.

Kalidas follows Kautilya and Manu's tradition on the relationship between the king and his ministers. In Malavikagnimitra, King Agnimitra used to hold council with his ministers in a secret chamber. The King Atithi, who was a friend of the pious people, was assisted by his wise ministers. These ministers personally settled complicated disputes between litigants in no time.[29]

Thus, ministers were regarded as the pillar of the state. In times of calamity, it was difficult for a state to survive if there was no support of the ministers. Kalidas agrees with Kautilya who states that when there is a fear of king's death, the ministers should declare that the king is engaged in a rite for removing the troubles of the country or destroying enemies or for getting a son.[30] Kalidas cites the example of Agnivarna who was bedridden and it was his ministers who proved themselves worthy by the performance of sacrifices to get a son.[31] When the issueless King Agnivarna died the ministers kept the news secret until they installed on the throne the pregnant queen of the late king. Like Kautilya, Kalidas, too, believed that in an unfortunate situation in the kingdom secrecy should be maintained otherwise there might be aggression by the enemy or revolt from within the kingdom by the seducible forces of the country.

Territory (Sovereignty)

Kalidas favoured a centralized administrative system under a king. He does not believe in division of the country into small conflicting independent principalities. Therefore, he loved to describe the monarchs of the solar races as the emperor[32] of the world and their empire stretched upto the ocean. He wished to establish an ideal kingdom ruled by a king who has controlled his senses, who is highly learned in the Vedas, Sastra and different arts and who could make his subjects happy and protect them from all dangers.

Treasury

According to Kalidas, treasury is important for the maintenance of the state. He states that by possession of treasure, the king becomes an object for patronage of others, just as the cloud with water in its interior is greeted by Cataka bird.[33] Kamandaka has also emphasised on prosperous treasury. To seek virtue (dharma) and material well-being, to pay servants and to overreach the calamity, the treasury should be guarded by the king.

Forts

According to Agnipurana, the invading monarch should carefully note that part of the citadel over which crows come and go and his concentration should be to attack that quarter. If this were done, it was believed that the forts should be captured in the end. Before starting on a military expedition, Kalidas says that the king should see that the capital fortress as well as the frontier fortifications were well defended.[34] The value of a fort to the king is incalculable. Only after such precautions had been made Raghu set out his famous digvijaya yatra.

Fear of foreign attack was not the only criteria for the construction of huge forts. Kalidas questions, "does not the Lion, the king of animals and a fierce destroyer of the gigantic elephant have its abode in the mountain cave?" Additionally, there are advantages of a fortified palace.

Danda

Kalidas appreciated adequately the principle of danda (punishment). The descendants of the solar dynasty punished the guilty in accordance with the crime so that the subjects can lead a peaceful life. Similarly, Machiavelli also expressed the same theme by stating that harsh punishment may be imposed on the offender in order to bring stability and security in the state.

A king was generally regarded as the supreme judge. Hence, his decision was final. King Raghu by applying the principle of adequate punishment was able to capture the heart of his subjects. King Aja took his judgment seat to look into the affairs of his subjects. In his very early age, King Sudarsana achieved supreme success of applying the principle of danda. About King Atithi, it is said, "he who had counselors for his associates and who was very vigilant, personally looked everyday into the affairs of litigations between plaintiff and defendants".[35] Even Yagnavalkya's opinion is important because the king had to personally look after the judicial affairs in the company of expert Brahmins.

The notion that a king should deal with danda proportionately has been stressed by Kalidas by giving importance to the solar kings. The aim of danda is the security of the subjects. If the people committing offences are not to be punished then the environment of the society will be affected. King Dilipa used to punish the criminals. A king should be just while delivering justice. For the king, severe rod became source of terror, while one with mild rod is ignored. The king who is just in the use of his rod is the right one. Raghu, like the southern wind, neither very cold, nor very hot, attracted the heart of all by just use of danda.[36]

Ally

According to Kalidas, "friend kept in low position are incapable of doing any favourable return, promoted friends become disturbing." King Atithi believed that a friendly disposed king should be kept in the middle position. It showed how a king should behave with his ally. Friendship was also made on the basis of marriage. King Raghu thought of the advantages of an alliance with Bhoja dynasty by marrying his son Aja with the daughter of Bhoja dynasty.

War

In ancient India, the concept of war was not something new. We have seen great kings, from time to time, used to indulge in battles to expand the territories. Generally it was seen that as soon as the battle was over, the king renounced the feeling of enmity. After the battle, the victorious king used to make efforts and secure the welfare of the subjects of the enemy's territories.

Out of the four classes, it was the kshatriya class that protected the kingdom by waging battles with the enemies. The training of the kshatriya warriors was based on instilling the spirit of pugnacity in them. Peace to them was a kind of activity until they did not subdue the whole earth. If there was no part of the country left to win this pugnacity was to be applied by him in protecting the subjects.

In Raghuvamsham, there are references of four great battles fought by various kings. These are: (i) Raghu's fights with different kings (Canto IV); (ii) Aja's fighting with the princes, gathered at the svayamvara ceremony of princess Indumati (Canto VIII); (iii) Rama's fights with Parasuram (Canto XI); and (iv) Rama's fights with Ravana (Canto XII).

In Raghuvamsham, Kalidas co-relates the heroic deeds with the conception of kingship. Battle is a natural corollary to kingship and the reign of a great king must be marked by dazzling military victories. According to Natyasastra conventions, battle scenes were not shown in the drama but they could be described in detail in poetry. Many epithets allude to the king as a great victor and yet military power is tempered with righteousness and restraint Raghu has been described as a righteous conqueror.[37]

Kinds of War

Ancient Indian thinkers have mentioned two kinds of war. These were dharmayuddha and Kutayuddha. Kautilya has mentioned tusniyudda as another form of war. Dharmayuddha was based on righteous principles of fighting battles while in Kutayuddha principles of dharmayuddha were scarified.[38] Kalidas preferred the just or righteous war but he did not also rule out the possibility of the use of deceitful means in exigency of circumstances. He cites the example of Atithi who was well versed in Kutayuddha[39] and who employed unfair means but for the sake of performing the Asvamedha's sacrifice.[40]

After the conclusion of battle, the victorious king secured land, money or ally as its reward. In the words of Kautilya, in between the gains of an ally, money and land, the gain of latter is preferable than the other two. For both the ally and money came from the gain of the land and ally came from the gain of money.[41] The simile used by Kalidas that the land becomes the source of treasure and the power of subjugation betrays the knowledge of such idea.

Related to these kinds of war are different kinds of kings engaged in warfare these are: (a) the righteous conqueror; (b) the greedy conqueror and the demonical conqueror. A righteous conqueror generally remained contended with the formal recognition of his overlordship. The greedy conqueror takes the territory of the defeated king and the demonical conqueror take the lives of the sons and wives of the defeated king.[42] In Raghuvamsham, there are references of sattva, rajas and tamas. Sattva included natural character, in born disposition, principle of vitality, truth, reality, certainty and the quality of goodness and purity that stands parmanant among the three gunas. Rajas meant passion, emotional darkness, quality of all material substances and activity seen in creatures. Tamas was marked by mental darkness, gloom, ignorance, illusion and error.[43]

In case of King Raghu, Kalidas said that the king who was conqueror for religious rites, seized the hoards and not the kingdom of the captured lord of the mount Mahendra whom he had then released.[44] Raghu victory was based on dharmavijai. He accepted the tributes in coins from the defeated kings but never seized their territories nor harassed their sons and wives. Moreover, Kalidas has depicted Raghu as a benevolent ruler who was secured after the victory. The acquisition of territory was of three kinds: new, formerly possessed and inherited. Raghu pacified the territory which he got either from his father or acquired it.

A Cakravartin King

In ancient India, most kings acquired the status of Cakravartin or digvijai king. According to Rapson, it was always regarded as a legitimate object of the ambitious king of every dynasty to attain the position of Cakravartin or supreme monarch. If the king is not eager to gain it, he falls short of an ideal king according to Rajadharma. King Dilip was a Chakravartin king who commanded the whole sea – girth earth, without any rival.[45] King Dasartha and Raghu were also Cakravartin kings.

Being a Cakravartin king, a king should watch the activities of the feudatories because any weakness of the king may enable the feudatories to strike and weaken the power. King Dasartha knew it well that the goddess of wealth (royal fortune) may depart through any loophole.[46]

Kalidas gives a detailed description of Raghu's digvijai. According to Pollock,[47] Raghu conquers the same territory as Samudragupta though he moves in the opposite direction. Actually, the Allahabad prasasti does not give us a clear picture of the precise directions and sequence of Samudragupta's military campaigns whereas Kalidas description of Raghu's expedition is a carefully constructed clockwise circumambulation of the subcontinent. Kalidas glorifies king Raghu by comparing him with Indra who turned east to conquer the world by threatening the enemies with certain defeat.[48] Raghu's unequalled reputation would reach his enemies in advance. It was followed by the deafening roar of the armies and behind the swirling dust one could see the chariots making the last column. In this sense, the armies moved in four different segments.[49] Like a mighty tusker jolting, uprooting or destroying trees that cause obstruction on the way, Raghu let some enemy kings go after extracting a tribute, killed several others in the battlefield or razed to the ground a few kingdoms.[50] In this way clearing the path he moved ahead.

Having conquered the eastern kingdoms, Raghu reached the shores of the ocean. The ruler of the Suhna kingdom saved their lives and their domain by accepting Raghu's suzerainty otherwise the latter would have uprooted the unyielding root and branch.[51] Raghu, then, defeated the kings of Bengal who had advanced to fight with armadas. Thereafter, he raised towers of victory all over the Gangasagar islands.[52] Later, Raghu and his forces crossed the river Kapisha. From here, he moved on to attack Kalinga. The ruler of Kalinga advanced with a fleet of elephants and attacked Raghu with missiles. After defeating him, Raghu's armies then covered a long distance and waiting for more conquests they landed on the Malayachal slopes. Here, the Pandya kings could not withstand the onslaught of Raghu's blazing might. The Pandya kings of the south offered a tribute to Raghu in the form of pearls at the confluence of Tamraparni and the ocean.

After this conquest, the valorous Raghu had a long sojourn in the Malaya and Dardur mountains where stood sandal trees on the slopes. Crossing the Sahyadri mountain, Raghu moved forward away from the sea. To move the ocean away from the Sahydari mountains, Raghu's armies marched westwards, along the mountain side. Fearing an onslaught of Raghu's forces many women in Kerala had run away. Then he moved towards west. A fierce battle was fought between Raghu's forces and the cavalries of the kings of the western country. Severing them by the strike of bhalla arrows, he littered the earth with the heads of the Yavana soldiers.[53] The enemy surrendered and Raghu marched forward with great valour; he defeated the Hun kings. Even the Kamboja kings could not resist Raghu in battle. They paid obeisance to Raghu. The defeated Kamboja kings offered Raghu several times a large member of horses of finest breed and enormous wealth.[54] Thereafter, Raghu's cavalry climbed atop the Himalaya. Here also fierce battle was fought between Raghu's forces and those of the kings of mountain. Raghu's armies shot arrow while the hill men hurled stones on them. With a barrage of arrow Raghu defeated Utsava and Sanket, the mountain princes. The mountain kings offered heaps of precious stones as tribute to Raghu. Having defeating them, Raghu turned homewards rather than proceeding to Kalilash. Having conquered them the victorious Raghu returned to Ayodhya. He performed the Vishwajit yagna on his return from the conquest of the world. Thus, Kalidas, while enumerating the achievements of these kings, was well aware of the concept of universal conquest that could make a king a Cakravartin king. He in his mahakavya tries to bring out the importance of victorious in battles so that the kingship and sovereignty could be properly established in the kingdom. In the conception of sovereignty and kingship, Kalidas can be compared to Kautilya. He depicts a more idealistic picture of a king through his imagination. His heroes are complete models of perfection and are worthy of imitation.

Season of War

In undertaking the expedition, the element of time, place and weather played an improvement part in ancient India. Kalidas has emphasized the importance of first battle by Raghu with different kings. The month of margasirsa (autumn) was generally recommended for an operation for war. In the tropics the rivers are flooded during the monsoon season and become fordable in the autumn. So the season of autumn was preferred and Raghu accordingly planned his operation in the autumn season.

Army

In ancient India, bala or army was considered as an important element of state. One cannot imagine war without proper arrangement of army or use of force to face the opponent. Kalidas too, emphasized this aspect of war. His Cakravartin king could not be a great king without a well-organized and disciplined army.

Kalidas, like Kautilya, mentions six kind of armies. There were maula, bhrta, sreni, suhrt dvisat and atavi. Besides this, Kalidas also mentions six types of forces, namely, elephant, horsemen, foot soldiers, chariots, spies and the guides. He sometimes distinguishes between foot soldiers and the infantry.[55] But the Agnipurana makes no distinction between infantry and the camp followers. According to it, the functions of the infantry consisted in carrying away the wounded and dead troops from the battlefield, offering resistance to elephants, supply water and carrying arms.[56] Kamandaka refers to six fold army and considers diplomacy and treasury to be its parts.[57] Later on, the importance of infantry was supported by Hieung Tsang who states that the mobility of the foot soldiers contributed to the defence of the kingdom.[58] From 4th century BC to 11th century AD the foot soldiers continued to account for the majority of the Hindu army.

Kalidas also speaks highly of cavalry. In Raghuvamsham, it is mentioned that after the defeat of the king of Kamboja Raghu received the best quality of horses. In the Gaudavaho Vakapati it is mentioned that the horses were the product of the Himalayan region. The reason for opting horses from foreign soil was the paucity of good horses within the kingdom and often compelled the powerful kings both in the north and south to receive the supply of horses from foreign soils.

So far, elephants were concerned, Kalidas, while describing the military operation of Raghu, mentioned mighty elephant forces of the king of Anga and Kalinga. Vakapati states that the King of Vanga was powerful because of the possession of a large number of war elephants.[59]

It seems that the navy was also an important element of war. Kalidas characterises the inhabitant of Vanga as expert in the act of plying boats and he clearly stated that Chandragupta II routed the forces of Vanga which were equipped with navy and implanted pillar of his fame on the waves of Ganga.[60] There is an indirect allusion to naval warfare in Kamandaka who says that by regular practice one becomes adept in fighting from chariots, horses, elephants and boats.[61]

Weapons

In ancient India, both offensive and defensive types of weapons were in practice. Dhanurveda classifies them as mukta, amukta, and yantramukta. The bows and arrows were the main weapons of the mukta type. The very fact that Indian military science was named Dhanur Veda sufficiently proves that bows and arrow were the principal weapons of war. Kalidas was well acquainted with the use of the bow is seen from Raghuvamsham. He refers to the twang of the bowstring of both Rama and Lakshmana that were headed by Tataka.[62] In Abhigyamshakuntalam there are several references where Dusyanta is credited with high skill in archery.[63] Thus, the chief weapon was bow which was drawn to the ear and not as Greek to the breast. The making of arrow is described in the Agnipurana as follows. They were made either of iron or bamboo. The latter was well grained, of good vein and of golden colour. Three kinds of arrow are distinguished, the best, the middling and the inferior. The shaft of the superior type of arrow measures twelve mustis (perhaps one musti was one pala), of the middling one eleven mustis and of the rest ten mustis. There is similar reference in Manusmriti. Kalidas mentions sharp, deep and strong shafts.[64] Kalidas also mentions venu, sara, salaka, dandasara and naraca. Spear was another weapon used by the solar kings. It was two cubits long and had a long and broad mouth in shape like the leave of a tree. Its end resembled the udder of a cow and were four hastas in length. It required the use of both the hands. The blade was of metal and the handle was made of wood. Kalidas puts its use for cutting the body. The weapon Isu is translated as a hand sword by Dr. Oppert but may be an arrow or a reed like weapon. Halayuddha has been identified as Kasastra.[65] It had a bent front and the whole measured two cubits in length. Four movements were associated with it. The mudgara[66] was a staff in the shape of a hammer which was three cubits in length and with a strong circular handle. The parigha[67] was a wooden beam used for looking or shutting the gate and it was a battering ram. According to Vaisampayama, it was made of wood and was circular in shape. It was so heavy that a whole army would be required to handle it. The sataghni[68] literally was one such weapon which had the power of killing hundred enemies at a time. It looked like a gada and was said to be four cubits in length. It had a strong handle and also possessed a number of iron spikes in it. It is generally identified with the modern cannon and hence was a projectile weapon of war. Mallintha quotes from Kesava in his commentary on the term occurring in the verse 95 of chapter 12 of Raghuvamsham:

> "*sataghnic tu catustala lohakantaka samcita yastih it Kesavah.*"

Ethics of War

Ethics in war means "to fight to the death for these great possessions (defence of family's honour and kingdom's welfare), to feel that it is a dishonour to die for them and dishonour to retreat from them, to give up every material for them, to fight for them to the last man, is just, true and righteous war and there is no thought or a passion connected with it which does not exalt and enoble, not only the persons and the nation which wage it but the whole body of humanity also to whom the great tradition of this defence is handed down and whom the memory of it inspires, teaches and ennobles.[69]

The Hindu science of warfare values both niti and sourya, i.e. ethical principles and valour. The term nitisastra shows that though the subjects treated in them were secular, still they had their bases in religion. In fact religion governed the actual life of every person from birth to death. It was, therefore, realised that the waging of war without regard to moral standard degraded the institution into mere animal ferocity.[70] Kalidas states a king desirous of dharmavijai should confirm to the code of ethics enjoined upon warriors. The king is not in favour of unscrupulous methods of warfare. Kalidas strongly condemns the policy of Kutyayuddha or war waged by secret methods. Though King Atithi was well versed and completely trained in the art of Kutayuddha,[71] but still he followed the principle of dharmavijai and because of his good qualities he was chosen as the favourite king by the royal glories of his opponents. He won the hearts of the enemies also. Kalidas describes the noble principles and relations of his heroes with their rival kings.

Foreign Policy and Diplomacy

Kalidas mentions the seven constituent elements of a kingdom in the very first Canto of the work. These constituent elements are the basis of the six measures of foreign policy. There are sandhi, vigraha, yana, asana, ashraya and dwaidhibhava. Sandhi means settlement against enemy to its fruitful end and uniting against a common enemy by making certain arrangements. Vigraha means war and hostility. Yana is marching against or attacking an enemy. Asana is maintaining a post, a cantonment against an enemy. Ashraya means providing succour to an enemy who has surrendered. Dwaidhibhava is double dealing, duplicity, deceit, keeping apparently friendly relations with an enemy, dividing one's army and encountering a superior enemy in detachments, harassing the enemy by attacking them in small hordes.

Kalidas's kings were well conversant with the six expedients of foreign policy. He states a king should neither be harsh nor too mild in the circle. He should adopt a middle course. King Raghu, who had all these qualities, made the adjoining princes bow down without extirpating them. A king should not be harsh towards the enemies if so, himself invites his own defeat. King Dasartha was also well aware of this. He was kind hearted to those who did nor transgress his command but was iron hearted to those who were hostile towards him.[72]

Kalidas places importance to Yana which is related to the marching and movement of the troops. The rule says that the king should find out the accurate estimate of his own strength, weakness and other requisites (time, place etc.). One should march upon the opponent if he finds himself stronger than the enemy, otherwise he should remain quiet and adopt the policy of asana. In other words, operation should be made against those who are likely to be overcome by the conqueror king. King Atithi followed this rule. Though he was possessed of power yet his operation was against those who were likely to be overcome by him.[73]

The idea of Vijigisu seems to have been invariably an king who is above all of the kingdoms in the circles. It was also seen in Raghuvamsham where Kalidas regarded Dasartha as the central king of a mandal of twelve kings.

Kalidas also mentions four upayas of war. These are sama, dama, danda and bheda. These are the four principles on which the policy of a kingdom is based. Sama means conciliation, pacification, negotiation, gentle and mild treatment of subjects. Dama is to restore, reconcile, give up, yield, surrender, entice, extend a privilege, grant and bestow. Danda means use of force and punish the offender. The spectre of a king is a symbol of his authority. Bheda mean splitting, opening a chasm, betraying, sowing dissension in the enemy camp in order to win over a faction.

These upayas are generally expressed in term of diplomacy. The Sanskrit term for diplomacy is naya. A king who understands the codes of diplomacy is sure to win the earth. King Dilipa was famous for his power of tolerance and spirit of forgiveness which is essential for a king being part of the upaya of sama. Rama, after defeating Ravana, installed Vibhisana on the throne of Lanka and thus won over him to his side. Kalidas states that political expedient applied at the right moment would get the desired result. Atithi, who was conversant with this, adopted the four methods of statecraft and gained uninterrupted results.[74] Moreover, to Atithi it was a sign of cowardice to depend on diplomacy alone. The use of brute force is like the wild animals that habitually fight and kill. However, he combined the two to overcome his enemies.[75]

Spy System

Kalidas had given sufficient importance to spy system in conducting the smooth functioning of the administration and protecting the kingdom from internal and external threats. Spies were generally appointed to gather secret information of his subjects as well of the enemies. The king's success depended upon the ability of the spies. Through them, the king was closer to his subject and thus, gained their loyalty. Kalidas says that both men and women were employed as spies.

Kalidas mentions the efficiency of work done by these spies when Rama was the ruler. Rama was able to know the views of the subjects through them. He asked the spy Bhadramukha[76] to report truly about the public opinion regarding him. As the sun illumines all corners of the sky with his rays and nothing remains hidden from him, similarly King Atithi had developed a network of spies in his kingdom that nothing relating to his subject was hidden from him.[77] Every day he discussed his state policy with his ministers but the policy was kept confidential. The policy was made public only when it was applied in practice. Moreover, King Atithi, in order to knew about his employees and his enemies, he very cleverly used his spies that even the spies were not aware of each other. Thus, successful implementation of foreign policy and diplomacy made the kingdom more secure and the subjects enjoyed the freedom of peace, prosperity and discipline.

Conclusion

Basically, Kalidas was more known for his shringara, viraha, karuna, vatsalya rasas of Sanskrit poetry, drama and literature. Yet his contribution on the military art cannot be denied also. He dealt in detail about kingship, the relationship between the king and subjects, the administration, war, foreign policy and diplomacy. He was well versed in the art of warfare. Kalidas wrote about a dynasty of kings. His kings were great chivalrous, mighty, sovereign ruler, unassuming warriors and supporter of gods. Moreover, they were shrewd in diplomacy and adept in politics.

An interesting feature of Kalidas' work was that he wrote about the people and for the people. This has been beautifully depicted in his work.

To him, man is the best creation. By gradual development he presented human characters at their best and free from all evils. In this, characterisation, Kalidas lays emphasis on ethical values of human life. The important question is if men were the centre of attraction how they lived their live peacefully. Kalidas, then, illustrates the greatness of the various king of solar dynasty in protecting the life of the people so that they can live peacefully. If this happens it happened during that time the people were freed from all greed, lust or any kind of evil, which could have brought the name of kings to shame. Thus, the primary duty of the kings were the welfare and protection of the kingdom by having an efficiency army. Kalidas places much importance to the military aspect in his various works. Kalidas has, from time to time, depicted Raghu as a great warrior by extending his kingdom is all directions. Much of what was written earlier is true even today.

In his writing, Kalidas owes much to Kautilya and Smritis. Inspite of their influence on Kalidas' writing one find that his work was a combination of realism and idealism. He wrote about the actual state of affairs in consultation with his ministers on all important matters but the poet's imagination laid much emphasis on idealistic choices and desirable practices. Inspite of all this, Kalidas's contribution can be seen in such a way that he gave a new flavour to his ideas so that they could appear to be his own original contribution.

REFERENCES

1. U. Singh, "The Power of a Poet; Kingship, Empire and War in Kalidas Raghuvamsham", *Indian Historical Review*, 38 (2) Sage Publication, New Delhi 2011, p. 178.
2. Raghuvamsham, 111, 13.
3. Ibid., IX. 18.
4. Ibid., VIII, 19.
5. Ibid., XVII, 63.
6. Ibid., IX., 6.
7. See R. Mishra, "A Critical Evaluation of the Theory and Practice of Kingship as Revealed in the Raghuvamsham", *The Journal of the Ganganath Jha Research Institute*, Vol. 23, March, 1967, pp. 113-46.
8. U. Singh, op.cit, p. 187.
9. Raghuvamshama, op.cit, 1. 19.
10. Kautilya, Arthasastra, BK. I. Ch. 6.
11. Raghuvamsham, XVII. 7. p. 57.
12. Manu, VII. 224.
13. Mahabharata Udyoga, 124. 37-38.

14. Kautilya, op.cit, BK. 1. Ch. 7.
15. Raghuvamsham, 1.25.
16. Ibid., 1.7.
17. Ibid., IX 8 and XVII. 42.
18. Ibid., II. 53.
19. Ibid., 1. 27.
20. Ibid., 111.20.
21. Ibid., 1.24.
22. Ibid., 1.32.
23. Ibid., IV. 12.
24. A.D. Singh, Kalidas, *A Critical Study*, Bhartiya Vidya Prakashan, Varanasi, p. 301.
25. Raghuvamsham, XIV. 23.
26. Ibid., IV. 87.
27. Ibid., VIII 17
28. Kautilya, op.cit, 1.7 p. 40.
29. R. Tandon, *Kalidas, The Raghuvamsham*, Rupa and Co., 2010, p. 759.
30. Kautilya, op.cit, BK. V. Ch. 6.
31. Raghuvamsham, XIX, 52.
32. Ibid., 1.5.
33. Ibid., XVII. 60.
34. Ibid., Canto IV 26, 17, 52.
35. Ibid., XVII. 39.
36. Ibid., IV. 8.
37. Ibid., 4.16.
38. Ibid., XII. 69.
39. Ibid., XVII, 69.
40. Ibid., XVII. 76.
41. Kautilya, op.cit, BK VII. Ch. 9.
42. Ibid,. BK XII, Ch. 1.
43. R. Tandon, op.cit, p. 864.
44. Raghuvamsham, IV. 43.
45. Ibid., 1.30.
46. Ibid., IX. 15.
47. S. Pollock, The Language of the Gods in the World of Men: Sanskrit, Culture and Power in Pre Modern India, New Delhi, 2007, p. 341.
48. R. Tandon, op.cit, p 208.
49. Ibid., p. 209.
50. Ibid., p. 211.
51. Ibid., p. 212.
52. Ibid.
53. Ibid., pp. 223-24.
54. Ibid., p. 227.
55. Raghuvamsham, IV. 36.
56. B.B. Mishra Polity in the Agni Purana, Calcutta, 1965, p. 167.
57. Kamandaka Nitisara XVIII. 2.
58. S. Beal, *Life of Hieung Tsang*, II, p. 139.
59. P.C. Chakravarti, Art of War in Ancient India, Dacca, 1941, p. 55.
60. G.P. Sinha, *Post Gupta Polity (AD 500-700)*, Punthi Pustak, Calcutta, 1972, p. 148.

61. Kamandaka, op.cit, p. 39.
62. Raghuvamsham, 11.15; 12.103.
63. Kalidas, Abhyigyansakuntalam, 11.4 and 6.
64. Ibid., 11.0.
65. Raghuvamsham, 12.23.
66. Ibid., 12.73.
67. Ibid., 12.73.
68. Ibid., 16.84.
69. See S.A. Brooke, A Discourse on War, the Hibbert Journal, 1915-16, p. 674.
70. Raghuvamsham, 17.48 and 68.
71. Ibid., XVII. 69.
72. Ibid., IX. 9.
73. Ibid., XVII, 56.
74. Ibid., XVII. 69.
75. R.Tandon, op.cit, p. 762.
76. Raghuvamsham, XIV, 31.
77. R. Tandon, op.cit, p. 763.

6

KAMANDAKA

Introduction

One of the important characteristics of ancient thinkers in India was that they laid much stress on state and government. This aspect could be seen from two perspectives. First the state was given a well defined role and second was the state of anarchy. In both the cases, the role of the king became important in providing a secure and prosperous state so that the people could be free from internal and external dangers. In such an environment, the new science of statecraft emerged and gained certain characteristics. The idea of Vijigisu, the state of the economy, the tendency towards kingships, the importance of war and its methods of implementing it in time of crisis and diplomacy as the instruments of state policy and the question of morality or immortality in the quest for realising the objective of the state were all influenced by prevailing social and political conditions.

The Arthasastra school arose in a period of grave political rivalry and continuous warfare. The kings became more and more ambitious to extend their imperial dominions. They were resorting to war and diplomacy and it depended upon the situation of the king in terms of strength and weaknesses. The Arthasastra thinkers also dealt with politics and conduct of these kings where moral considerations were subordinate to imperial aims. This imperial aim of the king could be achieved by well established armies which were prepared to wage a war for their imperialistic designs.

As such, Kamandaka's *Nitisara* is one source to establish political and military thought in ancient India. The contents have been analysed in collaboration with contemporary literary sources of his period. M.N. Dutt has stated that the essential characteristics of *Nitisara* are its gravity and sententious senses. The work influencing the later writings is borne out by the number of its commentaries as well as its has been quoted as an authority in later works on statecraft.[1]

There is much controversy about the date when Kamandaka wrote the book. M.Winternitz, R.C. Majumdar, D. Devahuti, D.D. Kosambi, A.L. Basham agree the work belongs to 8th century AD. But K.P. Jayaswal believes the book is written in 4th or 5th century while R. Shamasatri takes the period between 400 to 600 AD.

As the author of the *Nitisara*, Kamandaka's work is influenced by Kautilya's Arthasastra. The work is generally regarded as a summary of Arthasastra. Further Jha has placed greater emphasis on the comparative study between Arthasastra and *Nitisara*. In *Nitisara* the idea of kingship, state structure and governance, and methods to defend those institutions have been discussed in details. Honestly, Kamandaka regarded himself as a true discipline of Kautilya. *Nitisara* comprises 19 sections, 36 heads and 1163 maxims. According to Altekar, "this work is nothing but a metrical summary of the work of Kautilya". *Nitisara* does not describe the administrative machinery, the king and his courtiers dominated the political affairs of the state. Civil, criminal and personal law are untouched by Kamandaka because the Smriti writers of the period has already done great work on it. Kamandaka, like Kautilya, was more concerned with the material life of the people and the requirement of the state. One can says that Kamandaka's work is a summary of Arthasastra. Two-thirds of *Nitisara* is devoted to the causes and conduct of war and foreign policy.

Still then, *Nitisara* of Kamandaka may be reckoned as the representative of Arthasastra during this period. The author, in its own genuine style has dealt in detail with the acquisition and protection of the territory. Similarly, like other Hindu thinkers, Kamandaka has mentioned four kind of learning: Anvikshiki (logic), Trayi (Knowledge) of three Vedas), Varta (science of economics) and dandaniti (science of politics and art of government). Dandaniti has been analysed as the beginning of all other branches of learning. Self control or punishment is the name for danda. A king depends on danda and his policy is dandaniti. The king should protect himself with this and other branches of knowledge. Dandaniti is beneficial for all the people and the king is the protector.[2]

Implicit here is also the impact of danda on the subjects. Disobedience to a rightful master will produce unpleasant results which could be violent or coercive. It is not only dependent upon the master's physical resources but also dependent upon fate. When the king employs danda he may use force but if another man uses force in the same way to threaten a rightful king that would not be danda. It would be a sinful conduct of fallen man. Clearly much depends on the theme that the two rivals might both considers themselves rightful kings. But danda is more than physical thing. It meant the principles of justice and hence, the right to punish which was instructively accepted, at least as much as an act of violence.[3] This lead to another aspect of Kamandaka's thought, i.e. the nature and role of state and political institution in the form of kingship when the state is threatened by ambitious designs of other kings.

State

Kamandaka, like other ancient thinkers, conceives state in the form of having seven limbs (elements). The seven limbs are the king, minister, fort, treasure, people, army and friend. If any of the limbs fails, the state will be affected. All of them are related to each others. Kamandaka, like Kautilya mentions of the calamities of these elements affecting the security of the state in these words; 'Alongside this *Nitisara*, combines, however, incongruously, a notion that was lightly dealt in Asrthasastra, i.e. the concept of the organic relations of the factors of sovereignty. Kamandaka applies in one place the epithet which is helpful to these to seven limbs and he explains this by saying that sovereignty does not flourish even if one single limb has to face certain problems. In this passage what is evident is an idea of political theory which would be called the integration of the government units.[4]

Regarding the functions of the elements of the state are concerned, Kamandaka says that the king's assertion and the formulation of policy will well serve the interest of the state. If he is afflicted with calamities, he causes its ruin. In one of the extract (VI-23) he, after dividing the state into two units, the king and the territory (rashtra)-forming its inner and outer parts respectively, stresses that these two should be regarded as one because of their interdependence. Nevertheless, he envisages that all the other units of the state owe their origin from the rashtra and enjoins upon the king accordingly to improve the rashtra in all ways.[5] Kamandaka makes a distinction between rashtra and rajya. Rashtra stands for janpada while the traditionally categorised seven elements of a kingdom constitute the rajya. Like Kautilya, Kamandaka views the state as a means for preserving order and prosperity. How state is preserved? It is because of some political control, i.e. the king, could protect the state in times of internal and external crisis.

The King

Kamandaka considers the king as the most vital organ of the state. He is more necessary than the rains, for the failure of the rains state can be survived but the absence of the rulers cannot. This thinking is also true in the modern age. Without a sovereign government functioning of the state is affected. The king who wishes to keep a government in good conditions must study well its nature. Kamandaka says that injury to the king is more calamitous than injury to the rajya itself for the king is the active agent who is able to cure calamities that befall the state. Hence, a flourishing king cannot well obtain his objective by worthiness and one who is qualified is fit to wield sceptre.[6]

Nitisara mentions that the royal prosperity is difficult to attain and more so to retain and which entirely depends on the good will of the multitude rests steadily only on moral purity like water in a vessel.[7] A king should possesses the following qualities: nobility of birth, equanimity, youthfulness, good character, benevolence activity consistency, veracity, respectful behaviour towards elders, gratefulness, good fate, keen reasoning power, relationship with the great, ability to conquer his enemies, unshaken reverence, farsightedness, energy, purity, ambitious aims, modesty and piety.[8] The qualities will be appreciated by the subjects and the king will work with more vigour in maintaining and securing the state.

Furthermore, Kamandaka speaks of other indispensable qualifications of a king. These are eloquence, self-confidence, accuracy of memory, stateliness of stature, superior might, self-control, ingenuity for inventing various means and instruments of torture, perfection in all the arts, ability of easily reclaiming men treading evil ways, the power of sustaining an assault of the enemy, knowledge of all the remedies against danger, promptness in detecting the weak points of enemy, familiarity with the nature of war and peace, strict observance of secrecy regarding all consultations and actions, proficiency in turning into account the advantages of place and time, collection of money from the people and its proper expenditure, a deep insight into the nature of the dependents, free from anger, covetousness, fear, malice, obstinacy and fickleness, avoidance of tyranny, depravity, animosity, jealousy and falsehood, amiable appearance and appreciation of worth in other people. The king with such qualities make the subject happy and prosperous as they are living under the roof of their own father. Kamandaka calls such a sovereign Parthiva, i.e. a ruler of the earth.

A kingdom increases in prosperity which has a Mahendra (Indra) like king. Such a king is adorned with these royal qualities and whose acts are just and impartial. Mahendra is the god of firmament, the regent of the atmosphere and his world is called savage.

Kamandaka almost reproduces some of the basic ideas of his older masters. The importance of the king is seen from the subjects perspective because the king is there to protect the people from internal and external dangers. Protection also depends upon the king and science of agriculture, cattle breeding and trade. Like the clouds the king is the refuge of all creatures; if the clouds were to go wrong, the creature could still live but they could not survive if the king were to go wrong. According to this, the king's office is the primary as well as the essential condition of existence.[9] He also shows how the happiness as well as the misery of the people depend upon the personality of the king. One thing is clear that Kamandaka does not indulge himself in propagating the theory of divine right of kingship. But, the king following the dharma will enjoy the kingdom for longer period. Dharma alone will bring prosperity to the kingdom and to the king. As such, if the people honour the king even as they honour Prajapati (Brahma), then the virtuous king will protect his subjects well and conquers the kingdom of the enemies.[10] Thus, one can say that the purpose of the king is the protection of the people. Both are equally important to each other.

Government

Since the king is the key store of the government, scope of its activities was determined by him. In actual terms, the duties and functions of the king, as prescribed by great ancient Indian thinkers, constituted the scope and nature of governmental activities. According to Kamandaka, the territory and power of the state are extended qualities of the earth. So the king inorder to increase his wealth and objects of enjoyment, the land must have qualities like producing in large measures agricultural and mineral wealth. It should also have wealth of cows, enough resources of water, plenty of land and water routes which are capable of producing enough grains even without rain. Such land is meant to increase the means of desirable enjoyments. All efforts must be made to increase its wealth and happiness because this is bound to increase the virtues and pleasures of all its people. In addition to making the people happy and prosperous, the government should also perform all other functions necessary for maintaining law and order, administering justice and protection of the people.[11]

Minister

The king exercises executive, legislative and judicial powers of the government. But the king cannot perform all these functions alone. Hence, ministers or Sachivas provide source of strength to the king. According to Kamandaka, minister and crown prince are the arms of the king and the ministers are his eyes: in the absence of any of these the king becomes invalid.[12] *Nitisara* says that the council of ministers should be held by the king unwatched by others in the royal place at a spot having no pillars, windows, clefts or anything that might harbour an eaves-dropper or in forest. According to Kautilya, the agenda of the council comprised: (i) means of commencing operations; (ii) providing man and materials; (iii) distribution of place and time; (iv) counter action of disaster; and (v) successful accomplishment.

Kamandaka also mentions two points: one which is implied in Kautilya passages while the other is not mentioned at all. The first is that an item of the council agenda should be discussed more and more before its final disposal. The second is that they matter already resolved upon in the council should again be deliberated upon by the king himself in order that all flaws may be removed therefrom.[13] He also speaks of war council which possessed great powers.

Kamandaka also mentions about the role of the minister in advising the king. A king inclined to evil deeds should be prevented by them. He should regards their advice in the same light as those of his spiritual guide.[14] The monarchy collapses with the collapse of the king, while it revives with his revival, like the lotus reviving at the rising of the sun. Therefore, ministers imbued with genius, energy and equanimity and devoted to the interests of their master, should instil knowledge in him in a suitable manner.[15] Furthermore, *Nitisara* mentions that those ministers only are considered to be true friends and spiritual guide of a king who deter him from going astray. Those who restrain a king inclined to evil deeds were to him most worshipful preceptors and were not friends. When a king is blinded by passions, pride and arrogance the exertion of his ministers serve to extricate him.[16]

Treasure

According to Kamandaka, the treasure (kosa) should be large to contain lot of wealth acquired through rightful methods, full of desirable commodities and served by people. The head of the treasury should be able to bear the burden of any expenditure that may arise. The treasurer should always protect the treasure for purposes of dharma and artha. The king is the root of the treasurer and the duty of the king is not to indulge in fraudulent activities but it is to be used for the welfare of the state. A king possessing treasure as well as army are respected by the subjects. The treasure is the root of the army and the army is the root of the treasure. It is by maintaining the army that the treasure and the kingdom may prosper and the enemy is destroyed. The king should not incur greater expenditure than income but should increase the revenue from agriculture, trade, mines, etc. But the king should spend the same at the time of need for promoting dharma, artha and kama and if for the sake of dharma the treasure is reduced there is no harm in it.[17]

Danda

According to Kamandaka, the king protects his subjects by punishing those who deserve punishment. Punishment should be proper and rational. Proper punishment, given in accordance to policy, is the cause of wealth. Absence of punishment is the name for the law of the fish. Without use of force, the earth through under pleasure and great sink into hell. As an invalid sick and poor husband gets a virtuous wife through danda, so the subjects for fear of danda observe laws. Punishment given in a violation of niti is not correct but only proper punishment is praiseworthy.[18]

The Art of War

The ancient Arthasastra scholars opine that strong defence system of a state is the sign of a vigorous and powerful kingdom. In the absence of such a system a state cannot last long. So, to make the sovereignty of a king or a state unassailable and long lasting, the king should build up an impregnable defence system. Army is considered to be one of the vital elements of the state. All the seven elements of the state are inter-related with each others but it is defence system or army that becomes important when the state is faced with internal and external threats.

Before discussing the importance, role and organization of army, it is essential to discuss the concept of war. Kamandaka does not define it but enumerates the various causes of war. To him, wrath and resentment are not the only causes that breed war but a desire for evaluating one's position or excessive oppression by the enemy, usurpation of the kingdom, abduction of the women, seizure of the provinces and portion of territory, carrying away vehicles and treasures, arrogance and molestation of dominion may lead to war.[19] We do not find them mentioned in Kautilya's Arthasastra.

Kamandaka mentions the means for extinguishing the wars caused by the above factors by those skilled in the expedients of policy. The means of pacifying the wars caused by the violation of laws should be restored and the property to be protected effectively. When the king's kingdom is attacked by the enemy then the king is to adopt the same measure in defeating the enemy. Wars caused by taking away the treasures or fought by allies and imposing oppressive rule and persecution should be brought to an end by the restoration of the property taken and in case of allies behaviour, he should not join them but remain indifferent. In such a case none of the evils of the war may overtake him. It is always the duty of a sovereign to make common cause with his allies but not when he is faced with a war with him.

Furthermore, Kamandaka envisages a war resulting from insult, in that case, the king should be extinguished by the offer of honours. Here, conciliation is the proper mean for pacifying a war caused by pride and arrogance of one party. Therefore, a brave king should wage a war which is caused by the destruction of a friend or an ally. He should apply such measures by taking recourse to incantations and magical spells. In case, a war is caused by the eagerness to possess the same object, a prudent king should give up that object, provided that his royal prestige does not suffer.

Treasure is the wealth of the state. If the enemy takes the wealth of the state, then the king should wage a war and capture the lost wealth. Similarly, when the enemy is more powerful than a king, then the king show should sow dissensions in the enemy is camp by sama, dama, bheda and danda, Here, Kamandaka was not offering something new. These measures were well taken by his predecessors.

Kamandaka says that the hostilities may be of five kinds; (i) that produce by a spirit of rivalry; (ii) that caused by some dispute about land; (iii) that having women at its root; (iv) that produced by irresponsible spies; (v) and that consequent on some fault or transgression on one side. The last one is the ordinary kind of hostility, which originates from some guilt or transgressions.[20] The son of Valgudanti (Indra) speaks of four kinds of hostilities only; (i) that caused by the invasion of one's territories; (ii) that caused by something (done by others) prejudicial to the exercise of the regal powers; (iii) that results from some disputes about the boundaries of dominions; and (iv) that produced by some disturbance of the mandala.[21] Furthermore, *Nitisara* says that men have to take cognizance of two kinds of hostilities; (i) that which is hereditary and (ii) that bred by some fault or transgression. The second one is common sort of hostility which is bred by offer of insult and offences.

These are the sixteen kinds of war that should not be launched upon and tenaciously adhered to by a prudent king. A wise king should wage only such a war from which advantages may be derived both at present and in the future.[22] By accomplishing such acts, productive of present and future good, a king never brings shame on himself. When war has already knocked at his door, a politic king should pacify it by means of the expedients of policy. Victory in war is a thing uncertain and, therefore, one should not suddenly fall upon another.[23]

Kamandaka says that when an intelligent king finds his own army happy and efficient than that of his foes in the reverse state then he may launch upon war. Here too, he was not offering something new. Even today, as it was in ancient time, it is not prudent to enter into war with an enemy without a strong army.

Hence, army of a kingdom should be streamlined and strengthened and to make the military system more pragmatic and flawless, all the organs of the state should be well-knitted and well-equipped. Army (bala), then, becomes important because it is the prime duty of the king to protect its boundaries of the kingdom and people from external aggression and also to prepare for war against the enemy king, if ordered to do so by the king.

In order to protect the state from the enemy and for the prosperity of his subjects, the king should take all the necessary measures to recruit people in army, arrange for proper training facilities for them and also provide them with well-equipped arms and ammunitions.

Classification of Army

Most of the ancient scholars mention six kinds of soldiers viz., maula, bhrtaka/bhtra, sreni, mitra, amitra and atavi/atavika. Both Kautilya and Kamandaka also mentions the recruitment of the soldiers are classified into six categories. These are: (i) maula or hereditary appointed soldiers; (ii) Bhrtaka/Bhrta or the hired, paid regular army; (iii) Sreni-bala or the suhrd-bala (Kamandaka) or the martial troops of the king's allies; (iv) amitra bala (Kautilya) or Dvisadbala (Kamandaka) or the slave like soldiers recruited from the enemy camp; and (vi) atavi bala (Kautilya) or Atavika bala (Kamandaka) or soldiers recruited from jungle tribes. Another class of soldiers is autsahika. The soldiers of this category come from different regions and they remain always prepared to voluntarily attack any enemy kingdom whether asked or unasked by the king.

About the wings of the army, there is no unanimity amongst the scholars. Generally, the existence of four divisions of the army are accepted by the scholars. Manu, however, refers to six divisions of army – Kamandaka and the author of Agnipurana also mention six divisions of army. These are infantry, cavalry, elephant, charioteer, mantra and Kosa.

Kamandaka, while emphasising the importance of cavalry, states that the work of the cavalry is to reconstruct the forest and roads, to protect transport and supply, to cover the rear of retreating army, to execute a work quickly, to chase the retreating soldiers of his enemy and to kill them in thousands. Though Kamandaka refers to the function of cavalry in brief, there is no major difference between these two literary texts.[24] The Vishudharmottara Purana states that the cavalry is the chief limb of the army and is the best conveyance for covering long distances.[25] Somadeva Suri in the later period says that the cavalry represents the mobility of the army. With a strong army war becomes almost a sport and even enemies at distance easily come within his grip.

So far the role of elephant was concerned the task was to penetrate and break the forest forts. Both Kautilya and Kamandaka thought that the victory of the kings in battles were depended mainly upon elephants. Kamandaka, on this aspect gives more detailed account than his predecessor by stating that the kingdom of the king depends upon elephants and that one elephant duly equipped and trained in the methods of war is capable of killing 6000 well caparisoned horses. Later on, with the Turkish invasion, the Turkish leaders appreciated the utility of elephants in war.

Thus, we find that there is harmony of thought on the division of army as mentioned by the ancient smritikaras and Kautilya and Kamandaka as well as Ramayana pertaining in building a formidable defence system of the state in order to protect it from internal disturbances and external aggression.

Kamandaka mentions the qualities of these different types of troops. Maula troops are more reliable than the mercenary because for their respect and love for the king and helping in times of crisis. The mercenary troops are again more reliable than the Sreni troops because mercenary troops depend on the king for their livelihood. The Sreni troops are more reliable than the troops of the allies because the allies do not enjoy a share of the king's victory whereas the sreni troops participate in his joy and grief, and moreover live in the same country with the king. The allied troops are more reliable than the troops of the enemy weaned over, in as much as the former peruse the same object with the king and their kingdom and times of action are known whereas the latter are faithless, greedy and sinful. Moreover, they are wild and undisciplined. Hence this kind of troops cannot be trusted. A king has great reason for apprehension about forest tribes and the weaned troops of the enemy for secret overtures that may be directed towards them by the enemy. In such a case, the king, like them, should indulge in intriguing because intrigue will help the king in winning over the enemy.

Most of the ancient scholars have established the supremacy of maula of hereditary class of soldiers because they are more reliable as they have been tested and proved their loyalty to the king from time to time. According to Kamandaka, the maula army should consist mainly of the ksatriyas. Like Kautilya, Kamandaka mentions the occasions when the various types of soldiers should be mobilised.

Kamandaka mentions of a king whose resources are small for fighting a battle but by his power of counsel would make the allied troops like his own and thus the strength of the troops may be increased. He says that when the intention of both the king and the allied king are same then both jointly will have to show clemency and cleverness. When a king is supported by a large portion of the enemy's troops weaned over, a king should march against a powerful enemy. His position will be similar to a dog who is waiting to kill a boar. In such a case, he should bring into action conciliation or other kinds of policy. Kamandaka says that the forest troops should be employed in similar task and when entering into the territory of another, a good king should always place them in the front. A king who creates breaches in the enemy troops and who undertake difficult acts can detect and remove the cause of alarm of his own troops. A king possessing all these characteristics will make a great leader. Under him, the army is duly protected from evil influences.

Leadership

In expedition, a king should know all the features of a country, languages and human characters. He should also decipher all writings and possesses a relative memory and keen intelligence that would ultimately lead to success. A king who knows the times of sunset and sunrise routes, direction and the position of the stars and the planets and their consequent influence, is placed in a favourable position against the enemy. Similarly, a king will be in better position when he is neither frightened nor fatigued by the pangs of hunger and thirst and the condition of the weather.[26]

Battle Order

Kamandaka says that a king will move with his army arranged in due order after carefully assessing the rivers, mountains, forests and difficult regions of the kingdom. As such, the guide supported by a detachment of heroic troops should march in van and his camp and the treasury should be in the middle. The horses should march in both the flanks and they should be flanked by the chariots warriors. The chariots warrior should be flanked by elephants supported by the forest tribes.[27] The king after placing the troops in the front should march slowly towards the rear.

When the king feels danger in the van, the troops should be disposed in the Makara (crocodile array) or in the two winged Syena (hawked array) or in the Suchi (needle array) and then march forward. When the king feels that there is a danger in the rear, the sakata (chariots shaped) array should be formed. When the threat comes from the front vajra array should be formed. In all the situations, sarvatobhadra array should be formed when enemy is frightened.

Kamandaka says that a king should protect his troops when the troops are tired because of long marches, when they are harassed by robbers and distressed with diseases, want of food and oppression, when they fall asleep or become busy in preparing their meal and when they are not in proper ground and are not prepared for attack.

Kamandaka further says with the added advantages of the season and the soil in his favour, a king should fight a pitched battle otherwise he should fight in underhand ways. In unfair warfare, when the enemy is busy in pitching tents on unfavourable grounds, the king, who enjoys favourable ground should suppress the enemy. When a king is fighting on his own grounds the conditions are favourable to him. A king by placing the greater part of the army in the enemy's back should kill the enemy from the front. On the other hand, if the ground in the front is unfavourable, a swift moving king should kill the enemy from behind. A king should also suppress his enemy by going over his side who foolishly believes that he has conquered him. A king should also allure the enemy troops out of their camps, villages and forts and the then crush them.

Kamandaka envisages that a king should conceal the loss-efficient portion of the army and with the rest, supported by the allies, crush the enemy forces. If the remaining troops of the enemy are hidden the king should crush the enemy be enticing him away with the hope of plunder and then blockading his route of return. The troops that could not sleep through sheer fear of being attacked in the night and that have been worn out should be annihilated on the next day, Moreover, a king should take a lead by attacking in night with his fourth part of his army against the sleeping enemy. Kamandaka, thus says the seven kinds of hiding places are mists, darkness, herds of kine, pits, hillocks, under woods and river beds. Therefore, Kamandaka says a preserving king exerting in the right manner should crush his enemies by different kinds of warfare. The enemy movement could be known through the kings spies.

Camp

Kamandaka discusses in detail about the king's camp. The king while marching into the vicinity of the enemy's town, should be well acquainted with the ways of encamping. The camp should be quadrangular with four entrances. It should not be either too spacious nor too narrow and should be surrounded by all sides with highway, bulwarks and entrenchments.[28] The pavilion inside the camp should be made of square, crescent shaped, circular or long according to the advantages of the ground.[29] The pavilion should be such that it should be protected by powerful and veteran troops. A king must place near his pavilion the other types of troops in proper order – i.e., the old soldiers serving the royal line for generations, the rank and file, the troops of the allies and of the weaned enemy's troops and number of foresters in successive order.

So far the outskirts of the camp is concerned *Nitisara* mentions the king must place numerous formidable hunters of wicked deeds. They are paid well and are placed in circular array. Moreover, at the vicinity of the king's pavilion, fleets of elephants and horses under able dependents should be kept for protection. The king, for his own safety, must be armed with weapons and prepared for action during day and might.

The king, in order to surprise the enemy, should attack the enemy with a portion of his own troops along with allied troops outside his own camp. The duty of horsemen should ascertain the movements of the enemy troops. There should be strict vigilance all over by his faithful troops. Kamandaka says that the king should leave the ground which are spacious for the drill and exercise of his own swordsmen and sometimes he should destroy all other lands outside his entrenchments for the purpose of destroying the enemy's troops. The ground around the camp should be pervaded at places with thorny branches of trees, iron pointed pegs (caltrops) and with holes making difficult for the enemy to attack.

The king should choose such ground for the drilling of the troops which may prove advantage to him and disadvantage to the enemy. Kamandaka says the worst place is such spacious grounds which are good for the drilling of the enemy's troops. The king should try for the best encamping ground. If this is not possible then the king should select a middle one but cautions the king not to select the worst place inorder to gain success in battle.

Kamandaka says that the king should not camp at such places which are within the clutches of somebody, where numerous diseases prevail, which is blown over by unfavourable winds, where drums do not sound well, where there are constant alarms and fear, which is infected by flocks of crows, vulture and other birds of ominous note and where headless trunks of the tree are seen in the sun.

A camp should be such where people are cheerful, where drums and kettledrums sound aloud where horses neigh deeply and elephants duly equipped roar tremendously, where there is not dust storm, where favourable winds blow, where the troops are cheerful and where the asaras are in highly prosperous state.

Kamandaka, further, elaborates that when there are good and

auspicious signs in the camp the enemy force are easily defeated. But, in the reverse case the king had to suffer huge losses. The king will be victorious when he is supported by allies, wealth, knowledge, prowess, fortune, perseverance and manly efforts. The king is called Skandha because it is said that he is the root of the people. The functions of the ministers, the army and other members of the government are called Abara. Hence, *Nitisara* mentions, for the advancement and prosperity of the people, the Skandha or the king is supported by great Abaras or ministers or armies. This is known as Skandhabara. Therefore, the Skandhabara should be duly protected against the enemy.

Thus, proper encampment of the king's troops will place him in an advantageous position. In such a suitable situation, a king should march and take action against the enemy.

March

Kamandaka further, discusses the movement of the king's troops when it is on march. In all the marches, the king should first enter into forests and difficult tracts and to construct roads and passages where there are none. The king should try to break the enemy by his own united ranks. The functions of the elephant is to ward off sources of danger, to break down walls and gates, to protect the treasury and the uniform adherence to the policy from all dangers.[30]

While marching, the king should take note of forests, the different directions and the routes so that the supplies of food and the supporting troops are protected. The functions of the horses are to approach the enemy site and help the distressed portion of the vyuha or array. The functions of the infantry are always to be armed with weapons. *Nitisara* mentions that to clear of the enemy pits and passages, roads and tents and to know of the stock of fodder and food everything should be done in a similar manner as Viswakarma use to do.

All these are the qualities of infantry, cavalry, car warriors and horses. They possess good marks and follow all rules of conduct and Kamandaka says those who possess all these qualities will be employed for action.

Nitisara also describes the movements of various troops. The movement of the infantry is carried out properly where the ground is free from stakes and thorns and which possess outlets for retreat. For cavalry such grounds are good where there are small number of trees and stones, having no pits, creepers and caves and which is steady and free from mud and possess outlet for retreat. For chariots that ground is good for movement which is devoid of gardens and chasms and are steady and can bear the wheels. The grounds for the chariots, the horses and the elephant should be steady and hard. The wise king will never consider that ground suitable for the horses will also be suitable to elephants.

Nitisara mentions an intelligent king who is discourse of achieving victory should never indulge in fighting without a good cause. When the king is on the move, he should be placed on elephant and is guarded by lighter troops. The treasures should be carried where the king goes because royalty depends on treasures.

Kamandaka describes how the king should place the powerful forces in battle array. The number of horses should be three times the number of cars and elephants and five and five should be employed together. Infantry should be employed with them at the internal of one and horses at the internal of three. Elephants and cars should be placed at the internal of five. This kind of division is commended by all masters of polity.[31] The horses, men, warriors and elephants should fight in such a way that there efforts, in case of retreat, may be unobstructed.

Kamandaka further says that a king, surrounded by these heroic and brave persons, should stay on the battlefield and should fight inseparably and defend one another. The fiercest should be employed in the fight against the enemy. A good general is said to be the soul of the battle and it is lost if the there is no general.[32]

Forts

While considering the defence system of a state in ancient India, construction of forts were considered an important feature of art of war. The idea of forts, were not something new to Kamandaka. Descriptions found in Valmiki Ramayana show that strong fortification systems was developed and properly maintained by individual states in ancient India. Later on, Manu places forts at the third place amongst the saptang theory of a state. He refers to six kinds of forts. Sukracharya mentions nine kinds of forts and Kautilya states that there are four kinds of forts. All these have been discussed in detail in their respective chapters

Kamandaka says that a king should settle in such type of fort that has an extensive area and is surrounded by a wide ditch and secured with gates strengthened with high and massive walls, and which is protected by mighty mountains, forests and deserts.[33] These forts would stand in the way of an enemy who is coming to attack the king.

The king should have a full proof against the inclemencies of the weather, well stuffed with provisions and wealth and provision of abundant supply of water. *Nitisara* mentions without a fort a king is just like patches of clouds before a strong wind. Inside the fort, there should be natural hillocks and a large number of trees, under the cover of which the garrison at the time of siege would be able to give battle to the enemy. The site of the fort should be in a desert soil so that the besiegers would be compelled to give up their attempt out of sheer want of food and water.

Thus, a kingdom which has communication both by land and water and well furnished with castles afford better asylum place to the royal family at the time of siege. With such kind of arrangement, a kingdom is suitable for the habitation for the king who wants the prosperity and security of his kingdom.

Inter-State Relations

The Theory of Mandala

Supported by treasury and army, assisted by his ministers and officers of state and securing his fort, a central king should direct his attention to the consolidation of his kingdom.[34] Kamandaka, thus, envisages the concept of mandala. It ordinarily, means a circle. Mandala, in politics, signifies the circle of a king's near and distant royal neighbours. The number of enemies and allies included inside the circle varies according to different scholars. Generally, 12 kings form a mandala, namely, the Vijigisu wants to consolidate his kingdom by obtaining victory over his royal neighbours, the five kings whose dominions are in the front and the four kings whose dominions are in the rear of his kingdom, the madhyama and the udasina kings.

Kamandaka enumerates the qualities of Vijigisu. These are nobility of extraction, serving of the elders, energetic, ambitiousness in aims, power of penetrating into the hearts of others, keenness of intelligence, boldness, truthfulness, expeditiousness, generosity, humility, self-reliance, knowledge of proper place and time, resoluteness, skilfulness, physical strength, secrecy of counsels, consistency, courageousness and gratefulness.[35]

A king should regard those among his royal contemporaries to be his enemies who endeavour to appropriate to themselves the dominions and wealth on which he himself has set his eyes. On the question of Mandala Kamandaka quotes the opinions of different authorities. Mandala contains three hundred and twenty four kings. Of these, the circles marked 1.2.3.4.5.6.7.8.9.10.11 and 12 are of importance.

The circle marked 2 represents the dominions of the Ari or the enemy (of the Vijigisu). The contiguity of these two dominions are causes of conflict regarding the boundaries. The circle marked 3 represents the dominion of the mitram (ally). The dominions of the king of the 3rd circle are adjacent to those of the sovereign of the 2nd circle and before they become hostile to each other. The King of the 1st circle or the Vijigisu, takes the advantage of this hostility and befriends ari.

The circle marked 4 represents the dominions of the arimitram or the enemy's ally and hence he is enemy of Vijigisu. The circle marked 5 represents the dominions of the mitramitram who is friendly to the vijigisu and his allies. The circle marked 6 represents the kingdom of arimitramtitam or the ally of the enemy's ally. Hence, he has unfriendly relations with the Vijigisu and his allies. The circle marked 7 represents the kingdom of the madhyama (intermediate king). This king may become either an ally or an enemy to either of them.

The circles marked 8.9.10 and 11 signify the kings situated in the rear of the Vijigisu. The circle marked 8 represents the kingdom of the Parshnigraha or one who is situated in the rear. More often, this king is enemy of the Vijigisu. The circle marked 9 represents the dominions of the akranda or the king whose kingdom lies, in the rear, next but one to that of the Vigigisu. Akranda is one such king who prevents an ally from aiding another. If the parshnigraha has inimical relations with Vijigisu,

Akranda prevents the ally of the parhinigraha to join him and this he can do easily as his kingdom lies between parshnigraga and has ally.

The circles marked 10 and 11 represent the kingdoms of the two asaras or the two kings who respectively support parshnigraha and the akranda. They are, hence, called parshnigrahasaras and akrandasara. The circle marked 12 represents the kingdom of the udasina (neutral king). His kingdom lies beyond those of the enemies and allies of the Vijigisu. These twelve kings ordinarily constitute a mandala. The description of mandala has also been given by Kullukabhatta and Mallinatha in their respective commentaries on Manu.

These four, namely, Vijigisu, ari, madhyama and udasina are said to be the principal components of a mandala. Though various kinds of mandala have been mentioned by scholars but the mandala consisting of twelve kings is universally known and accepted.

The king is a real political who knows a tree, having four roots eight branches, sixty leaves, two props, six flowers and three fruits.[36] Kamandaka, then compares a mandala to a tree. The four roots, are Vijigisu, ari, madhyama and udasina. The eight branches of a tree are an enemy and an ally of each of four cardinal kings. The five prakritis of each of the twelve Kings of a mandala constitute the sixty leaves of the three. The six expedients to be used in foreign policy constitute the flowers. The result of the application of these expedition, namely, diminution, preservation and aggrandisement of the territorial wealth of a kingdom, are the fruits.

Thus, Kamandaka says a king, treading the path of justice and bringing about the purification of the mandala with all endeavours, shines resplendent like the autumnal moon of pure beams, affording delight to the hearts of the people.[37]

Adoption of Sixfold Policy

The six fold policy or Sadguna is know as the collection of six indispensable measures which should be initiated by the king willing to conquer over the adversaries. Like Kautilya, Kamandaka follows the six fold policy and discusses in detail about the various aspect of the sixfold polices. These are sandhi, vigraha, yana, asana, dvaidhibhav and samsraya.

1. **Sandhi:** A king, who is faced with powerful king and his kingdom is in danger, should seek for peace or make treaty with him. Kamandaka enumerates sixteen different kinds of peace. These are kapala, upharanas, santanas and sangatas; upanyasas, pratikaras, samyogas, purushantaras, adrishtanaras, adishtas, atmamisha, upagrahas, parikrayas, uchchinnas, paribushanas and skandhopanayas.

Kapala-sandhi is concluded between two parties of equal resources. Upharanas sandi is concluded through the offer of gifts. Santana sandhi is that which is concluded by a king by a giving a daughter in marriage to his royal adversary. Sangatas sandhi is founded on friendship between good men. This kind of peace is good like the gold among the metals. Upanyasa sandhi is one such treaty where peace is concluded with the aim of bringing into a successful termination of all the controversies on all occasions. Pratikara sandhi is concluded through benefits conferred and received. For example an alliance that concluded between Rama and Sugriva. Samyoga sandhi is that peace when two parties join one other for accomplishing an act that is equally important to them and they enjoy each other's confidence. Purushantara sandhi is to secure when the weaker party surrenders its troops to the use of the stronger. Adrista-purusha is a compound formed of adrista and purusha man. In such kind of sandhi, some of the conqueror's men are bound to help those of the conquered when the latter called upon by the former goes to perform some act for the former's benefit. Adista sandhi is such peace made with a powerful enemy by giving certain portion of territories. Atmamisha sandhi is the compact that is formed between a king and his own troops. Upagraha sandhi is concluded for preservation of one by the surrender of everything else. Parikraya sandhi is concluded through the surrender of a part or the whole of the treasure. Uchchinna sandhi (destructive peace) is concluded by the cession of most of the excellent lands to the enemy. Paribhusana sandhi is that which is concluded by giving up the products of the whole territory. Shandhopanaya sandhi is that which the vanquished party is required to carry what the conqueror may demand of money.

Thus, Kamandaka says that a powerful king invading another's territory does so with the objective to obtain good booty and unless he is offered valuable presents he will not give up his attempt. It is the upahera sandhi that force the weaker party to offer those presents. Thus, upahara is the kind of peace that is ordinarily concluded and is superior to all other kinds of peace.

2. Vigraha is declaring war against an enemy king.

3. Yana: Kamandaka refers to yana. It means the expedition undertaken by powerful and energetic king and whose subjects are loyally attached to him though his many excellent qualities. The objects of military expedition is to obtain victory. *Nitisara* mentions the word energetic or 'ulkristavala viryyasya'. It means the king whose army is in an excellent condition and whose prowess is great will obtain the object of war.

Eminent scholars speak of five different kinds of Yana. These are vigrikya, sandhaya, sambhuya, prasanga and upeksha. When a king marches forward for crushing his host of enemies by the sheer dint of his prowess, it is called vigrihya yana.[38] When a king supported, by his own allies, marches forth for complete crushing of the allies of his enemies by force, it is called vigrihya-gamana.[39]

When after concluding a treaty with the enemy in the rear the Vijigisu, out of a desire for victory, advances upon another enemy, it is called sanahaya-gamana. When a king, along with faithful, warlike and powerful samantas, marches against a common enemy it is known as sambhuya-gamana. When kings like Suryya and Hanumat, jointly undertook an expedition against an enemy that threatened the safety of the parkritis of both it is called Sambhuya-yana. When a king is victorious over samantas by promising rewards in case of success and then moves against his enemies, it is called sambhuya gamana. When a king originally marching against a particular enemy, afterwards through some contingency, proceeds against another, it is called prasanga-yana. Prasanga means some event or cause. What Kamandaka means to say that a king sets out on a march against a particular enemy, but owing to some incidents on his way, he changes his original intention and proceeds against another, although at the time of setting out he had no intention. Kamandaka gives the example of king Salya, the ruler of Madras who was the maternal uncle of Pandavas. As it is natural, the battle between Pandavas and Kurus, he set out to join Yudhisthira. But on his way, he was charmed by Duryodhana's skill and as a result he fought on his behalf. But, during the fight Salya was killed by Yudhisthira.

Kamandaka further states that when a powerful king marches against an enemy and has every chances of winning but the king disregards them and proceeds to fight against enemy friends. This is called upeksha-yana. It means that a king at first marches against a certain enemy of his and been successful, but without making the best of the occasion and completely crushing the enemy, he attacks the enemy's friends, who were off guard all the time satisfied with the victory obtained.

4. **Asana:** Kamandaka describes the term asana. When the king stops for a while either in the beginning or during the progress in war, it is called asana. There are five kinds of asanas.[40]

The first kind of asana is known as vigrihyasana. This happens when both ari and Vijigisu endeavour to thwart the plan of operation of the other. When again a king besieges an enemy it is also called Vigrihyasana. When an enemy is secured within the fort and becomes difficult to capture him, a king, then, should cut of the asara by blocking the road. The strength of the enemy is also reduced when his line of communication within the country is cut off. In this way, a king gradually subjugates his enemy.

The second type of asana is sandhyasanam. This situation arises when both ari and the Vijigisu had suffered a lot in war. The war ends by the conclusion of an armistice. Kamandaka gives the example of Ravana had to take recourse to sandhyasanam by offering Brahma as his hostage when he had to fight with the Nivatakavachas.

The third one is called sambhuyasanam. It means a king thanks that udasina and madhyama are equal in strength should await and muster all is forces inorder to make an attack on either of them. Here mention may be made of the term Sanghadharman. The uvayari (another name of madhyama) desires the destruction of the ari and Vijigisu because being more powerful than them he should withstand by his aim.

The fourth one is called prasangasana. This situation arises when a king wants to go to a certain place halts there through some contingency or other, at a place different from where he intended to go at first.

The last one is called upekshasana. This is caused by the indifferent attitude of a king before an enemy who is more powerful than him. Kamandaka illustrates this by citing indifferent attitude of Indra in case of parijata. Parijata is the name of one of the five trees obtained by the churching of ocean. The allusion is: On one occasion Narada, the celestial sage, while wandering through heaven was honoured by Indra with a garland of the parijata flowers. He gave this garland to Krishna who gave it to Rukmini who was then near him. Narada, known for brewing quarrels went to Krishna's second wife Satyabhama and narrated the incident to her by saying that she had been shamefully neglected by her husband by not giving the garland to her. When Krishna came be known about it, he promised to fetch for her the parijata tree itself from heaven. He requested Indra but he did not pay any attention to it. A battle was thereupon fought and afterwards through the mediation of other gods, the parijata tree was allowed to be transplanted into the garden of Satyabhama and Indra was advised to treat that act of Krishna with indifference.

5. **Daidhibhava:** The term is made of two words: daidha meaning double and bhava is the state or condition. Daidhibhava is of two kinds: swantantra and paratantra. Swantantra daidhibhava means the duplicity of a king who for effecting his own safety has taken recourse to it but this is done out of his free and independent will and not under any kind of pressure. Paratantra daidhibhava means dependents. Hence, this kind of daidhibhava is the double dealing done by spy practices and are commissioned by their employer. This kind does not give any direct personal benefit to them. Hence, one can say that daidhibhava of an independent person is called swatantra and that of a dependent person is known as paratantra.

6. **Samsraya:** Kamandaka describes another mode of policy which is called samsraya or seeking protection to a more powerful king. But there is a distinction between sandhi and samsraya. When a weak king is compelled to make an alliance with a stronger one, inorder to save himself from the persecution of a strong enemy, it is called samsraya. On the other hand, sandhi can also be concluded between two contending parties, the assailed and the assailant.

Thus, Kamandaka shows the inevitability and indispensability of these policies in the interests of smooth and efficient functioning of the state administration. A prudent application of these sixfold measures are conducive to the stability, security, development and progress of the kingdom. These policies serve the purposes of fulfilling defence requirements and diplomatic needs in a meaningful manner.

Diplomacy

Role of Ambassador

Kamandaka, like other scholars of ancient India, places much importance on the role of ambassador in the conduct of state's affairs. An ambassador deserves to be such person who is dauntless, accurate in memory, eloquent, accomplished in arms and in the sastras and capable of doing all sorts of works. Ambassadors are said to be of three kinds, viz. (i) those invested with full powers; (ii) those with restricted powers and (iii) those merely who repeat what they are instructed by their masters. In respect of rank, the last one is inferior to the former because they are invested with minor powers.

Kamandaka says that an ambassador should befriend the frontier as well as the forest tribes and should find out the inland and navigable communication and easy routes. This is done to facilitation the unimpeded progress and march of his master's army. An ambassador should not like spy correctly enter an enemies dominions but would enter with his open permission. He should not lie even when his life is threatened. It is his duty to deliver the message he has been entrusted with without slightest mutilation. When questioned on the subject by the enemy's king he should not speak anything about the weakness of the Prakritis of his own king but should in flattering language say "your majesty knows everything well."[41] The ambassador should praise the enemy's king by saying that he is on way in inferior to his own Vijigisu. After coming into contact with the treacherous element of the enemy's state, he should with their help know the former's movement so that the enemy could be easily weaned over. Sometimes he should be in touch with his own secret agents remaining disguised as ascetics in pursuit of knowledge of help places. He should place before the enemy his own master manliness, high lineage prosperity, forbearance, great energy, magnanimity and gentleness.[42]

Kamandaka, further says that an intelligent ambassador should not be depressed and hopeless regarding the accomplishment of his projects, although he would have to wait for a considerable length of time. In other words, delay should not deter an intelligent messenger. He should make the best use of that time by studying the condition of the enemy's kingdom in various ways. During the progress of any of these undertakings, the ambassador has every chance of getting hold of something likely to be beneficial to the informations gathered by him would turned out to be beneficial to the interest of his own master. If the ambassador finds flaw in the enemy he should not despair but content himself with the thought that even the enemy himself may march one day against his own king then he finds the enemy deferring everything to a future period, he should seek an explanation whether or not the enemy is gaining time to deprive his rival of the advantages of a seasonable attack. When once the proper time is gone, the enemy will have to wait long for another such opportunity. Thus, Kamandaka says once the task is accomplished he should directly come back to his own kingdom or remain there inorder to communicate to his king all the important points of his information.

Espionage Network

The secret agents also play an indispensable role in the security of the kingdom. A king desirous to secure victory against the enemy should develop an efficient and effective network for collecting information about the conditions prevailing in the enemy's state, its employees and also about its motive as well as the strength and weaknesses of the enemy's king so that he can take adequate measures against it. Hence, the role of spies become important.

Kamandaka discusses, then, the qualities of a spy. A spy, is such a person who is skilled in the interpretation of internal sentiments by conjecture and by external gestures, accurate memory, polite and soft in speech, agile in movements, capable of bearing up with all sorts of difficulties, ready-witted and expert in everything. Sly spies disguised themselves as ascetics and traders or artisans acts like wandering in all directions. They are well informed in everything. The spies are the medium through which a king perceives things that being away from his sight, cannot be directly perceived by him. Even when the king sleeps yet his spies are wide awaken and are working so that when he wakes up he will come to know everything important that may have transpired during his sleep.

According to Kamandaka spies are of two main categories: (i) visible envoys and (ii) secret i.e. secret watchers. Guided by his spies, a king should proceed to any work. The spy service is said to be well established when the ambassador is kept well-informed by the spies. Through them, a king should know those enemies who have reasons to be exasperated with him and also by secret measures punish him if found guilty. Those who have reasons to be exasperated with the king should be conciliated by gifts and honours. He, further, says that cool headed spies should study the conduct through signs and gestures, bodily appearances, the secret token they use and the letters they write. Thus, spies well versed in all acts and in all the arts should roam in every part of a mandala imbibing public opinion like "the solar rays imbibe moisture from the earth."[43] Thus, the information gathered by the secret agents helps the king to strengthen his position according to the needs and circumstances. In this way, Kamandaka has forcefully put forth his discerning thoughts on the defence system of a kingdom by protecting his kingdom and the subjects from external dangers.

Conclusion

Kamandaka's period was generally associated with the decline of the Kusan power in northern India and the Satvahanas in Deccan during the third century AD. The disintegration of these powers gave rise to new powers in north India under Chandragupta I in 320 AD. By the fifth century AD, the disintegration of the Gupta empire has also started. It was, then, Kamandaka formulated his political and military theories under a powerful king who could not only protect the frontiers of the kingdom but also bring happiness and prosperity to the kingdom.

In his writings, Kamandaka followed the footstep of Kautilya in propagating the political and military ideas which are still relevant today. In Kamandaka's *Nitisara*, they are certain portions which are not given undue importance. On the other hand, his focus on the theory of kingship, duties of kings, saptang theory and the theory of war and peace have been well taken and can be compared to any strategic writer. Infact, while explaining war and peace he was more elaborative than his mentor, Kautilya.

Kamandaka's greatest contribution was to agree that military strategy should be subordinated to political decision. In fact, this idea found supreme expression in modern age with the writings of Machiavelli,

Clausewitz and even today where war or military is subordinated to politics or policy. Somewhat like the writing of Carl von Clausewitz, the Prussian thinker, Kamandaka says that politics or policy should occupy the dominant position while conducting a war. The *Nitisara* criticises militarism and considers over dependence on military strength is dangerous. Military power should always be subjugated to political judgment. Furthermore, Kamandaka focussed both on just and unjust war depending on the king's relations with the enemies. But, before waging war, the king should be in a position to known the internal situation of the kingdom. If things are favourable the king should avoid the war. As such, he advocated the theory of paternal despotism. It means the king's attitude towards the subjects should be like father. In order words, he should try to win the people's 'hearts and minds'. In such a case, the king should use force or army carefully. But, a real threat to the kingdom from internal and external sources then force becomes necessary to bring stability and security in the kingdom.

Kamandaka's contribution in the realm of military lies in explaining in detail the concept of peace in detail. On propagating this idea, he went ahead of Kautilya by mentioning sixteen types of peace that has to be achieved by the king when faced with dangers. Today, Kamandaka's military writings influence the functioning of governments in maintaining relations amongst the states. Before, a king plans a military operation against his foe he has to carefully study the foe's intentions and this Kamandaka makes it quite clear by emphasing the role of ambassador, espionage system or even the mandala system. His emphasis on these systems could be well compared to modern system where state indulges into alliances or make peace for its security. Infact, in modern age diplomacy and espionage are the tools in the implementation of foreign and security policies before going for actual combat.

The best policy is to follow the law of collective and harmonious existence. Through this, Kamandaka was emphasizing the defensive nature of war. If defensive measures fail then the king should opt for aggressive action against the enemy. Again, this ancient idea of defensive-offensive action is even true with every generation. The only difference is that the development of science and technology, with the passage of time, has made the ancient strategy of war and peace more complex and destructive. No nation, today, therefore, wants war but wants to live peacefully, knowing the reality that the humanity or civilization will be extinguished by the use of deadly and sophisticated weapons. It would result in utter destruction. Hence, like *Kautilaya Kamandaka* military writings are still important in shaping the state's security policy.

REFERENCES

1. F. Jha, Comparative History of Kautilya Arthasastra and Kamandakaiya Nitisara, p. 46.
2. Kamandakiya Nitisara, II, 2-16.
3. I.W. Mabbett, Truth, Myth and Politics in Ancient India, p. 81.
4. U.N. Ghoshal, A History of Indian Political Ideas, Delhi, 1966, p. 216.
5. Ibid., p. 374.
6. M.N. Dutt, Kamandakiya Nitisara, Chowkhamba Sanskrit Series Office, Varanasi, 1979, p. 30.
7. Ibid., p. 31.
8. Ibid.
9. Ibid., pp 32-33.
10. U.N. Ghosal, op.cit, p. 37.
11. Kamandakiya Nitisara, IV, pp 50-56.
12. N.N. Law, Aspects of Ancient India Polity, p. 33.
13. M.N. Dutt, op.cit, p. 37.
14. Ibid., p. 38.
15. Ibid.
16. Kamandakiya Nitisara, V, pp. 61-88.
17. Ibid., 1-12; 37-43 and XIV, 12-13.
18. M.N. Dutt, op.cit, pp 36-37.
19. Ibid., p.141.
20. Ibid., p. 141-142
21. Ibid., p. 144
22. Ibid., p. 146
23. Kamandakiya Nitisara, XVIII, pp. 13-15.
24. Ibid, XIX, pp 4-5.
25. B.B. Mishra, Polity in Agni Purana, pp. 170-71.
26. M.N. Dutt, op.cit, 242.
27. Ibid., 243.
28. Ibid., 245.
29. Ibid., p. 246.
30. Ibid., p. 226.
31. Ibid.
32. Ibid., p. 249.
33. Ibid., p. 252.
34. Ibid., p. 253.
35. Ibid., p. 40.
36. Ibid., p. 81.

37. Ibid., p. 83.
38. Ibid., p. 93.
39. Ibid., p. 104.
40. Ibid., p. 150.
41. Ibid., p. 155.
42. Ibid., p. 81.
43. Ibid., p. 83.

7

BANABHATTA

Introduction

Banabhatta, one of the most romantic poets of his time, was well aware of the political and military developments in ancient time. His period was a period of expansion of state's territories. It demonstrated that the king's design was imperialistic in nature. The idea of imperialism was quite reflected in the writings of Kautilya, Kamandaka, Sukracharya and others. Besides this, like them, Banabhatta has mentioned the ability of the king to protect his kingdom from various sources so that the king could run the administration smoothly. Good administration would bring prosperity and happiness in the state and to Banabhatta it depended upon wealth and army.

Banabhatta is well known for his two meaningful books, *Harshacharita* and *Kadambri*. *Harshacharita* was written in praise of his patron king Harsha of Thaenswar (606-648 AD). The author has identified his work under the title of akhyayika. The other book, *Kadambari*, was written in the form of katha. The term akhyayika and katha signify a composition dealing with historical facts and is work of imagination respectively. Before Bana, Panini too had mentioned akhyayika. Panini quotes three akhyayikas by name, Vasavadatha, the Sumanottara and Bhaimarthi.[1] These compositions were in the form of prose and signify historical personages. But, it appears that akhyayikas had gone out of fashion after Bana's time. History testifies that there have been no work of this kind of later period than Harshacharita. From this perspective, in its literacy presentation, Harshacharita, then occupies a unique position in ancient India historical literature. Bana must be credited for writing the biography of King Harsha. Later on, Bana's work became a source of inspiration to later writers to depict the political and military achievements of Harsha.

Bana's other popular book *Kadambari* is regarded as the best prose romance in Sanskrit. It is written in Katha form. The book enshrines the imaginative and highly ethereal romances of the princes of Vidisa and Ujjani. The environment is a political one which visualizes the geographical and territorial frontiers of the northern states. The well known discourse in the book in the form of Sukanasopadea deals with the dark side of politics. The author's notion of politics and war could be well co-related by the narrative account of the king Sudraka and Tarapida.

Reading both the books one can say that Banabhatta was fully aware of the political, social, economic and military conditions of his time. He has laid much stress on state's affairs, society, religion and military organization. To the student of the art of war, the military observations of Banabhatta could be seen by the performance of the king in protecting the kingdom from internal and external dangers. It is in this light, the above factors become important in understanding his view on war and statecraft.

Society

The importance of society in ancient India was well understood by earlier writers like Manu, Kautilya, Kamandaka, Sukracharya and others. The society was marked by various distinct features and one of the features was its composition. All the writers had adequately dwelt on this point. Similarly, Banabhatta, considers society as comprised of four classes, namely brahmanas, kshatriyas, vaishyas, sudras and other mixed and occupational classes. In ancient India, the class composition worked as an ideal factor in systematization of the Hindu way of life. It formed a strong bondage in ancient social organization and was also instrumental in the preservation, exposition, continuation and popularization of Indian social, cultural, philosophical, moral, aesthetical and spiritual norms of life. From Kashmir to Kanyakumari and from Kashmir to Gujarat, the predominance of class structure has remained undisturbed and unchanged throughout the ages.[2]

Like Manu, Bana in his Harshacharita had systemically characterized the character of Harsha as one who "carried all the rules for the varnas and ashramas."[3]

This indicates that the state as such was secular in nature where everyone had the right to practice their own religion. Complete tolerance resulted in the happiness among the subject and which was a cause of prosperity of the king and his kingdom. The different cults like Buddhism, Jainism, Brahamanism, Lokyatas and others found proper place during Harsha's regime. In other words, the king was impartial towards them. It is interesting to note that Prabhakaravardhana was a devotee of the sun.[4] Rajyavardana believed in Buddhism and Harshavardhana was a staunch believer of Siva. With this kind of secular society, the people were well contended and satisfied that the king could protect them from all evils and threats that endangered the state.

Among the caste composition, the duty of protecting the society was given to the ksatriyas class. This was not something new. In ancient India this was a normal practice and this aspect had been well taken by earlier ancient Indian writers. Even the Chinese writer Yuang Chuang says that ksatriyas class was "the eye of the kings" and has held sovereignty for any generations.[5] Though Yuang Chuang mentions of other kings belonging to other castes but majority of them were ksatriyas. Bana had described two famous races for the ksatriyas. They were the Lunar and the Solar races. We do not find references in the Harshacharita about the existence of these races as the separate communities having their relations with the Moon and the Sun. But the inscription of the immediate post Harsha's period helps us, to a great extent, in reaching such conclusions.[6] Moreover, Bana says that the ksatriyas were noted for their patriotism, valour, courage and heroism. They were great warriors and fought many battles. The janapada of Sthanirsvara was regarded as the "land of the heroes by the sons of swords, i.e. ksatriyas.[7] They worshipped their weapons and it seems that it was the customary among the ksatriyas community.

State

The state in ancient India was manifested in the seven organs or elements which resembles to the four elements of the modern state. These seven elements were: (i) the king or the sovereign; (ii) ministers; (iii) janpada or territory; (iv) durg or fort; (v) kosa or treasury; (vi) bala or army; (vii) mitra or allies. In Bana's both books and even in Harsha's dramas these elements of state are well laid down.

The King

Bana has laid much emphasis on the divine character of the king. This could be well collaborated by the narrative account of the king Sudraka and Tarapida. Bana's observation about the king Sudraka reveals that he believes in divine dignity of the king. He says, "by reason of dharma dwelling in his mind, Yana in his wrath, Kubera in his favour, Fire in his splendour, Earth in his arms, Lakshmi in his glance, Sarasvati in his speech, Moon in his face, the Wind in his might, Brashpati in his wisdom, Cupid in his beauty and the Sun in his glory, Sudraka, resembles god Vishnu who manifests the diverse forms and who contain all the deities within himself."[8]

This shows that Banabhatta was in no way different to other ancient thinkers accepting the theory of divine right of kingship. References of divine right of kingship was seen in Manusmriti, Ramayana and the Mahabharata. The theory was practiced and became popular when the subjects were suppressed and the debate was focussed more on the issue of morality or immorality of statecraft. The idea of good or evil became more important when gods were harassed by the evils and they requested Vishnu to ordain someone who on account of his supreme powers may secure the highest place among men. Vishnu, then, appointed Virajas for the work. Prthu, the seventh generation of Virajas was the first king appointed by Vishnu that possessed the status of the king and no one could challenge his authority. Similarly, in Manu, the king is made up of divine elements drawn from various gods.

King Harsha was a benevolent ruler and his administrative ideal was the protection of the subject so that they could attain happiness and prosperity in the kingdom. In Kadambari and Harshacharita, Bana describes the king as the protector of his subject. He describes Tarapida as physically strong and enthusiastic king and was "the incarnation of justice and destroyer of the sorrow of his people." Tarapida, as a young king, had conquered the earth ringed by the seven dvipas by the might of his own arms which was as stout as the trunk of Indra's elephant.[9] Conquest of the world is taken as the symbol of imperial power and Tarapida attained this dignity even in his prime youth.[10] The primary duty of the king was to please the subjects. He made his subjects well contended even he had to crush his enemies. A king should also pursue dharma, artha and karma without being attached to anyone of them.

According to Bana, there are two kings: good and bad. A bad king is one who is ill-disciplined even if he possesses vast territory. Such kings get angry and punish the subjects and thereby cause harassment to them. A conceited king like this becomes troublesome like cupid.[11] The qualities of a good king is as follows. King Haraha's eyes are not stained by the deadly poison of pride. His speech is not choked by the disease of conceit. His postures are not such that he will forget his natural dignity by a sudden epileptic fit of forgetfulness caused by the heat of arrogance. His changes of feeling are not exaggerated by the burning fever of overbearing self-will. His gait is not agitated by the unnatural movements of an access of self-conceit. The jewels of his wisdom are attached to men of pure virtues. He takes interest in pearl like qualities and not in heaps of ornaments. He is inclined towards pre-eminent glory and not for the withering stubble of this life. His natural instinct is to help his friends. His sovereignty is to help his dependents. His success is an instrument of helping his king's folk. Power means to him helping the unfortunate and riches to help the brahmanas. The utility of his body is in carrying out the bold undertake. The earth is the arena for his sword, the circle of kings are needed only to amuse him and the enemies are to show his valour.[12]

From the above description, it is necessary for the king to be bereft of pride, self-conceitedness, arrogance, overbearing self-will and other shortcomings. He should honour the virtuous and meritorious, liberal in granting gifts and should exert to seek glory. He should show valour in subduing the earth honour the virtuous and meritorious and establish friendship with bow, help his companion and dependents, honour the learned, assist the brahmanas and should always be proud of his valour and never depend on his allies.

The King and His Subjects

For the security and prosperity of the state, it is essential that there should be stable relationship between the king and his subjects. No kingdom (state) could survive without the benevolent attitude towards his subjects. History is full of examples. In Kalidas's Abhijananasakuntalam, a king plays the role of a real father to his subjects. Towards his subjects, he behaves like a relative.[13] Bana echoes the same idea when Prabhakaravardhana at his death bed, consoles his son Harshavardhana with these words that for kings their subjects are the near and dear one and not their own relations.[14] Bana's view is quite different from Kalidas. In Shakuntala, the king held to be near relation to the subject and in Harshacharita the subject is considered as the kith and kin of the king. Thus, the relationship is one of intimate family relationship. Even Prabhakaravardhana consoles his son not to worry of his death because his subjects will support Harshavardhana as a relative to fill the vacuum created after the demise of his father. If this kind of relationship follows between the king and the subjects, the subjects will become more loyal and dedicated to the king. This is an important factor in enhancing the king's power and valour not only within the state but also among the states in the conduct of foreign affairs.

Ministers

In ancient India, generally, it was found that the king alone could never enjoy absolute powers. In administrative matters, the king had to rely on the advice of his ministers. In Mahabharata, it is said that "the kings are virtually depended upon ministers as animals are depended upon clouds, Brahmanas on the Vedas and women on their husbands."[15] Manu considered those kings "as foolish and unfit' who" could carry on administration by himself.[16] Kautilya too warns that "one wheel cannot move the chariot, the state cannot be administered without the help of others. The king can, therefore, only succeed if he acts in accordance with the wise council of ministers. The importance of ministers, purohits and other advisers were found in every age and it was seen that no king could afford to avoid the advise of them and become more autocratic.

Bana's description of the king is more interesting. He refers King Sudraka's ministers who were well versed in statecraft, which they had thoroughly mastered. They were free from greed and were affectionate and wise.[17] In Kadambari, the name Sukanasa appears as the prime minister of King Tarapida who was Brahmana by caste. In Kadambari, Bana refers to two prime ministers and both of them were brahmanas. These were Sukanasa, the prime minister of Tarapida and Kumarapalita, the prime minister of king Sudraka. They were the chiefs among the circle of all the royal counsellors. Thus, Bana was not propagating something new. Earlier, Manu and Kautilya had mentioned the prime minister was brahmin by caste. Bana mentions about the powers of the prime minister, Sukanasa. The king used to consult him in all administrative works. He was a profound student of art and sasatra. His strong affection for the king had grown up from childhood. His qualities included thoroughness in the art of statecraft, truthful, meritorious, good conduct and skilled in performing duties. Like the serpent, sesa, who endured the weight of the world, he was able to shoulder the responsibility of the government of the world. Like Jarasandha, he was expert in conducting war and peace. Like the ocean containing big animals, he was endowed with great moral strength. Like Shiva accompanied by Durga (Parvati) he had acquired durgas (fortresses). Like Yudhisthra, he was a day spring of dharma. He was the trusted counsellor of Tarapida as Brihaspati was of Indra, Sukra of Ursaparvan (the demon king), Vasistha of Dasartha, Visvamitra of Rama, Dhaumya of Yudhisthra and Dhamanaka of Nala.[18]

Similarly, in Matysa Purana it is mentioned that the king should appoint such assistants for himself who are brave, strong, endowed with wealth, beauty, nobility, self-controlled, capable of hard work, sweet speech, salutory advice, courageous and devoted to their masters.[19] Thus, Bana in Harshacharita observes that the king without pride is a truthful minister and a king's son without vices were difficult to find. With all these qualities of the ministers, it increased the king's power and prestige and subjects had to suffer less from any kind of threats.

Janpada (Teritory)

In ancient Indian literature we have come across various technical territorial terms which cannot be exactly be determined and made applicable to all periods and to all states. These terms were desa, rajya and rastra. So far desa is concerned, the stage manager in all the dramas of Harsha introduces to the audience the circle of kings coming from different countries as regions. In the dramas, Harsha uses rajya as a territorial term for a country. Bana, himself uses the word very clearly in the sense of a country. He uses the term "Parasikadesa" for the country of Persia. The terms janpada and janapadavisea used for Srikanta and Sthanisvara respectively. This may suggest that the country as a whole was known as desa, a particular region was known as janpada and the headquarter of the janpada was probably known as janapadavisea.

Kosa (Treasury)

Harsha's generous distribution of wealth at quinquennial assemblies held at Prayaga amply demonstrates that Harsha followed a sound fiscal policy. It does not mean that he accumulated wealth by taxation. This was largely due to his enlightened government.

Bana's description of kosa (Treasury) as an element of state is interesting. It is in the form of condemnation of Lakshmi (the goddess of wealth). This fills a large space in the canvas of Sukanasa's teachings. The poet has presented some general characteristics of Lakshmi such as her conduct with a king, her contradictory qualities and the evil consequences. Lakshmi is well known to all of us. Lakshmi, when acquired, is difficult to preserve. She disappears even after being fast bounded by the strong ropes of the six gunas (peace, war etc). She escapes even after being confined in the cage of swords brandished by a thousands fierce champions. She does not keep friendship, cares not the race, look not handsome, follows not the pattern of a family, never thinks for the disposition, counts not excellency, desires not to bear the sacred learning's, values not righteousness, honour nor liberality, counts not the power of discrimination, pursues not conduct, does not understand truth, follows not the men of auspicious marks. Being fickle by nature she makes no firm footing anywhere. Even when kept with so many efforts in palaces of the kings, she comes out as if being intoxicated by the excessive drinking of the ichor of their many wild elephants. She resides on the sword's edge as if to learn cruelty. She changes many forms. She renounces even a king is richly endowed with army, judicial power, treasury and territory. She resorts to the passionate people as a creeper resorts to the tree. Though producing wealth, she is unsteady by nature. Like Sun's rays she alights on one thing after another. She is full of dense darkness and can be won only by courageous man. She never comes to the learned and virtuous men, despises the great and gentle souls, "discards the man of noble family, avoids the brave, forget a liberal man, never reaches the modest. Though bringing the fever of pride caused by wealth she produces chill by destroying one's power of discrimination". Though rising from water she increases the thirst after wealth. She is a cause of jealously and struggle and finally result in woe. She enhances desires, attracts the senses, darkens the goods conduct, increase infatuation and pride, cover the eyes lighted by the knowledge of sastras, increases sins, hamper the virtues, encourages fraud and abolishes goodness and righteous.

Bana further mentions the adverse consequences of Lakhsmi. Sukanasa realizes that when this base evil creature Lakshmi wins the kings by the will of destiny. They being helpless becomes abode of shameful acts. Their graciousness is washed away. Their heart become black and the forgiveness is swept away. There is a class of kings who comes to be deceived by unsteady riches. These kings forget their origins in the arrogance of amassing a little wealth and are afflicted owing to the vehemence of passion increased by various misdeeds. They act frantically as if boiled by the heat of arrogance caused by wealth. They act crookedly like crabs. Their course of action is determined by the unrighteous. They, like lame persons, are unable to walk themselves and are thus led by others. On account of their pride, they do not recognize even their relatives. They cannot be looked at even the brightest virtue. Such persons cannot be roused to activity even excellent counsels. Bewildered by the poison of greed they see everything as golden. By their funeral pyre, the prosperity of these persons is extremely terrible.[21]

By focusing on the negative aspect of wealth, Bana warns the king to be more alert in conducting the affairs of the kingdom. He has shown the effect of wealth and power. In order to meet the four ends of life two important elements are artha (material means) and kama (desire). The poet does not lose sight of their importance for without adequate resources no individual, society or political institution can function and meet its ultimate goal. It has been pointed out that treasury is one of the seven elements of the state. Bana has condemned wealth not merely it is wealth but it affects an important person who lacks the virtue of discipline. Wealth gives birth to the sense of vanity and pride and sometimes acts as reckless arrogance in case of those who lack the strength of character and wisdom. The proper use of wealth is, therefore, for the well being of the state and its subjects. It is to be noted that Bana's negative approach towards wealth is a peculiar feature of Kadambari. Before Bana, several authors and poets have dealt with the beneficial aspects of wealth but none had approached it from the spirit in which he has ventured.

War

In any society or state there is a need for some kind of force that could bring the subjects to act as one unit. This problem arises only when a state is either threatened by external or internal dangers. In ancient India, there were small or big kingdoms which tried to maintain its independence through fair or unfair means. In other words, if these kingdoms became victim of attack they could opt for either just or unjust war (righteous or unrighteous war). It was seen that whenever a big kingdom emerged as a strong, and vast power, the small states became a part or paid tributary to it. The idea of supremacy was quite prevalent then as these states were constantly fighting among themselves. Supremacy, was based in the idea of unification when the big state failed to protect the smaller state or there was some kinds of internal or external threats. Then, these states used to rise for their own independence. In order to keep the state intact army was the chief instrument of defence and infact, it was regarded as one of the important elements of state.

So far Banabhatt is concerned, he lived in a period of political turmoil. It was assumed that there were some 70 kingdoms which were fighting for their existence. The survivalability of these states, to a great extent, was depended on waging battle. Bana does not define the term war or battle. But he says that in ancient India the kings or the princes took active part in the battlefield. They led their armies in the battlefields. In Harshacharita, Rajyavardhana after defeating the enemy was wrapped in "long white bandages bound about arrow wounds received in the battlefields.[22] It was, then, could be assumed that the battle was waged by two armed kingdoms for some specific objects. Primarily, it was the unification of its own kingdom and the other factor was expansion of its territories. Therefore, in battle a king or his ally has to prove his own prowess. A real valorous king is one whose powers is capable of securing the life of all the subjects and who never cares for the assistance of his allies. Bana confirms that the king used to go for universal conquest. Harsha, after becoming king made up his mind for the conquest of the four quarters. This could be possible only either through engagements or accepting his paramoutancy.

Proper Time of Attack

Bana says that it is the way of the vile that the king strikes their rival when they find a calamity befallen on him or when they find any loopholes in his kingdom.[23] This is what Samvadaka, the attendant of Rajyasree, remarks about the attack. To him, the people of vile character will only take the advantage of the weak points of the enemies. But, according to Kautilya, Kamandaka, Sukracharya and others, this is the normal feature in warfare. According to them, the ambitious king's aim is to find out the weak points of the enemy and attack him at the proper moment. That act is not vile. It is a political necessity. It was a compulsion for a kingdom to go for battle otherwise the kingdom would have to face internal or external threats. Kautilya has mentioned that when a king finds that his enemy's subjects are ill-treated, impoverished and greedy, oppressed by the thieves etc and so are easily seducible or while his own agriculture and commerce are flourishing while those of his enemies are suffering from famine, etc., then in view of inflicting injury on his enemy and for exhibiting his own power, he may proclaim war.[24] Kamandaka, too, subscribes the policy of Kautilya. In similar tone, Manu suggests that a king should never live by deceit, supports the idea that a king should inflict injury on his enemy at his weak points. Bana has also laid down his well known policy in Harshacharita. This was utilized by the king of Malwa when he found Grahavarman was helpless on account of the death of his father-in law, Prabhakarvarmana, he mercilessly attacked and cut him off from his life. At another place Bana mentions that after the death of Prabhakaravarmana, the king of Malwa found Rajyavardhana and Harshvardhana leaderless and then he contemplated to invade and seize Thansewar. This indicates that the enemy should be attacked if the king suffers from disability.

Offensive and Defensive Battle

In ancient India, it was a common practice that a princes used to participate in battle along with their fathers. But about Harsha's early military career, we do not have sufficient information. From Bana's Harshacharita, we learn after the marriage of Rajyasri, King Prabhakaravardhan one day summoned Rajvardhana to lead an expedition against the Hunas. He led a huge army comprising ancient advisers, and devoted feudatories towards the north to attack the Hunas. Bana, further, mentions Harsha used to march with the horses for several stages.[25] When Rajavardhana "had entered the snowy regions", Harshavardhana was said to have spent several days away from the camp where he used to get an opportunity to hunt lions, tigers and boars which were plentiful. He said "left the forest empty of these wild creatures in comparatively few days."

Bana further comments that kind of battle which Rajyavardhana waged was not merely an offensive battle. It was a combination of offence and defence against the Hunas. The battle was a serious engagement between the two warring parties. This indicates that Harsha's participation in the military operation makes it clear that Harsha was well trained for the battle and was allowed to go inorder to have practical training in military operations. Bana confirms that the king indulging in such kinds of military activities thought for universal conquest, i.e. the conquest of the four quarters. This could be possible only through offensive and defensive engagements or accepting his paramoutancy.

Indulgence in battles need certain amount of discipline, training and the application of proper strategy and tactics at appropriate time and place. All these qualities were inherent in Harsha. Banabhatt bears testimony to these facts. It, further, indicates that Harsha had maintained a huge army to face any kind of threats whether arising from within or from external sources.

Composition of Army

Generally, the king was the supreme authority of the state and was also the head of the army. As such, he took active interest in maintaining an efficient and disciplined army. Regarding the traditional composition of the army, the army consisted of foot, horses, chariots, and elephants. But Bana and Hiueng Tsang mention foot soldiers, cavalry and elephants and are silent on the use of chariots. Even the Mahabharata copper plate of Harsha mentions boats, elephants and horses but not chariots. In

Harshacharita Bana refers to another kind of force which is known as camel corps. Though at one place Hiueng Tsang did made mention of chariots. But for us, these chariots were not actually war chariots but was more concerned with the insignia of honour enjoyed by military officers. Then, there are references of chariots present in the battlefield and were surrounded by the foot soldiers. Again, the assumption is that they were not used in the battlefield and the foot soldiers only protected the life of the commanders and military who used the chariots. It seems that there was gradual decline of use of chariots duly equipped with followers and soldiers accoutred in mail and with the ranks of army of brave heroes, a monarch should march forward having consultations with ambassadors and the feudatory kings".[26] Similarly, Gandavaho of Vakapati also gives an interesting account of the military organization of Yasavardhana but is silent about his war chariots.[27] The Panchatantra mentions elephants, horses and infantry but not chariots.[28] Finally, Muhammad Bin Kasim who invaded and conquered Sindh early in the 8th century AD did not encounter war chariots in any battle which he fought.[29]

Regarding horses, Bana mentions that the royal stable of Harsha was filled with horses from Vanayu, Artta, Kamboja, Bharadvaja, Sindh and Persia. Even in Mahabharata there are references of horses brought from Kamboja and Gandhara which were the best horses.[30] The paucity of good horses in India often compelled the powerful kings, both in north and south, to get the supply of horses from foreign soil.

According to Banabhatta, the elephant corps formed the strongest division of Harsha's army. The Aihole inscription too refers to Harsha's mighty elephants. The Narmada battle may well be regarded as one of the most memorable battles from the point of view of displaying tactics in elephant warfare. Since the elephants formed the main force the king paid proper attention in raising this particular kind of force. Bana describes in details about his entrance to the royal gate. He saw many camps of elephants. He mentions twelve kinds of elephants which were kept under the custody of the camps. These were: (1) elephants brought for tying on their heads the fillets of honours; (2) to bear drum; (3) new bound; (4) acquired as tribute; (5) acquired as presents; (6) sent by the rangers of the elephant district; (7) brought in the honour of first visit to the court; (8) sent at the time of embassy; (9) presented by the lords of a wild settlement;

(9) let loose or kept ready for a watch of the elephant fight, engaged in fight or separated and led after attack; (11) the elephants that were placed at the watch – at the place where the hours of watch were marked and (12) collected together like so many mountains to construct a dam across the ocean in order to conquer all continents.

Elephant corps were not only maintained by Harsha but history is full of evidences where kings preferred to keep a contingent of elephant fleet. The reason behind it were many. As Kautilya says that they were the best mean of storming the enemy forts. They were also capable of enduring for sufficient time the rain of arrows showered by the enemies in the battlefields. Bana has mentioned about these advantages.[31] Because of the huge size, elephant in time of crisis, served the purpose of a bridge to cross over the river. Sometimes they served the purpose of watchers too. In view of ample utility of elephant in ancient warfare, we are not surprised that Bana's has devoted so much space to the description of elephant force. Bana's unusual description of Harsha's favourite elephant Darpa-sata as "a moving hill fort with its high frontal globes as so many towers", as an iron walled indented with thousand arrows" and as "time in his butting"[32] confirm the above qualities. With high wooden towers securely fastened to their backs the elephants were used as mobile fortresses. From the towers, the soldiers would manoeuvre their weapons, including nooses of various kinds.[33] The Sassanians too used their elephants in the same way. The reserve of the Sassanian army was formed of elephants brought from India which inspired the Romans with a certain amount of terror.

Bana also mentions about the management for the safety of these elephants. Elephants doctors were recruited to look after the diseased elephants. Forest chiefans were ordered to inform immediately if any herd of elephants have been seen.

Army Commander

The king was the supreme commander of all armed forces. For military administration, he appointed an army chief called senapati. Senapati was generally a man of great military vision, well versed in the art of war and enjoyed great respect at the royal court. Bana mentions Senapati Simhanada who occupied an important place at the royal court. He worked under

Prabhakaravardhana who treated him as a friend.[34] The senapati's work was to lead the army in the battlefields. He, according to Bana, was "the family priest of martial companies", the performer of proud speeches, the sustainer of the routed, the executor of pledges, the authority on opening in great battle and his very voice inspired the warriors with lust for battle.[35]

This graphic description of the senapati proves the armies were led by such generals and the king was more interested in their military victories. The important question is why Bana had referred to Senapati Simhanada and not Harsha's senapati. The answer could probably be that when Harsha ascended the throne Simhanada was quite old to take active part in military campaigns. But the King Harsha could have benefitted from his logical ideas and past military experiences. The duties of the senapati seemed to be performed by the mahasamdhivigrahedhikrita, the supreme minister of war and peace.

Under Senapati there were various military commanders performing the functions related to battle. One such was commander was mahabaladhikrita who was in charge of a particular corp. Bana further refers to baladhikrita which gives an impression that he was the chief of the barrack-superintendent (patipati). The chief officer of the cavalry was known as brihadasvavara.[36] It appears that he was also called asvapati. But the most important was gajasadhanadhikrita who was the commander of all elephant corps.[37] It seems that Harsha employed all possible means to acquire elephants. The list of officers employed in the elephant stables are numerous and it shows the importance of the use of elephant even in Harsha's time. The mahouts were also known as mahamatras.[38] They were skilled in training the elephants in various manoeuvres. There is also mention of adhoranas whose job was elephant riding and were trained to walk in different ways. The servants were called karpati[39] and lesika[40] who supplied the beasts with fodder.

Weapons

Bana mentions bow and bow strings, arrows, helmet, mace staves, lance staves, swords and sharp pointed spears.[41] Shields were used by the soldiers in their defence. Bana refers to kirdaranga shield of various colours.[42] His contemporary Hieung Tsang also speaks of large shields and large spears carried by the infantry vanguard of the army. In Harshacharita other weapons are also mentioned. Warriors riding on the elephants threw kamanda on the enemy's soldiers and brought them down easily. Hastipasakriti was also used in battles. By means of vagnra, the enemy's warriors riding on the elephants or horses can be brought on the floor and killed at no time. All these weapons were effective.[43] Kuntaka or nija was fastened to the bamboo of 12 or 15 feet in length. The warriors sitting on the elephants would use it effectively.[44] There is a reference of Vavanas in the Harshacharita. It means that Harsha was well acquainted with the aerial machines. The Tamil work Jivakacintamani refers to jivaka flying over the air.

March of the Army

The astrologers fixed the hour of marching suitable for the subjugation of the enemy from all quarters. Bana informs us that Harsha consulted a group of astrologers on the eve of his digvijya campaign.[45] Even in Chachnama it is mentioned that when Roychach prepared a grand army for an operation, he consulted several astrologers.[46] Kamandaka also advices that the king should worship the god and brahmanas on the eve of expedition.[47] When Prince Chandapida mounted and marched on the way host of kings with thousand elephants were swaying in confusion, their umbrellas broken by the pressure of the crowd, bowed low before him, their crest jewels falling low as the diadems, bent in homage, their erring hanging down and the jewels falling on their cheeks, bowing as a trusted general recited their names.[48]

In Harshacharita, we find a more realistic description of the manner of the movement of the army. In the early morning, when all creatures were slept and all was still, the making drum was beaten with a boom deep as the gaping roar of the sky elephants. Then, after a moment's pause, eight sharp strokes were distinctly given a new upon the drum, making up the number of the league in the day's march.

The commanders gathered crowds of barrack superintendents. Thousand of torches lighted by the people made in road upon the darkness of the night with their glare. Pairs were roused from sleep by the women guards which showed their presence in the camp. Shrill words of command from the marshal dispelled the slumbers of the blinkering riders. Elephants and horses were also awakened. The low people of the neighbourhood, running up as the elephants and horse has started looted moving. On the way they were heaps of abandoned grain. Donkey ridden by throngs of boys accompanied the march. Crowds of carts with creeking wheels occupied the trumpeted roads. Oxen were laden with utensils momentarily put upon them. This shows that the donkeys, oxen, carts and small horses for carrying loads in marching were provided by the poor villagers to the king.

When the march began the front of the king's residence became full of chieftans arriving from every side. They used to mount on female elephants with riders holding up bows stripped with gold leaf swords grasped by confidential servants occupying the inner seats. After a brief interval the king enters, riding upon a female elephant. The company of the feudatory kings greeted him and bowed with bodies dutifully bent down. In return, the king distributed his favour among them according to their position and prestige. He, then, reviewed the army and began to march. During the march, the attendants of the animals quickly cut with knife a mouthful of fodders from the beanfield. This shows that there was no protection of crops from the troops in the march during Harsha's time.

In the course of march, it seems that the army's movement was in a disorderly manner comprised of mass of soldiers through a friendly zone. The only precaution observed was to keep the king in the centre. The army on the march was accompanied by a host of women also. This included not only the king's harem and the wives and mistress of the great nobles and feudatory chiefs but also a great retinue of courtesans. In the end, the march presented a spectacle of gorgeous procession with banners unfurled, standards displayed and music galore. It was not something new and this practice of marching has been also mentioned by Kautilya.

Camp

Bana mentions that when the kings were not in the capital, a military camp made of bamboo and grass is put up.[49] But we find in the Kadambari that the military camp of Chandrapida was full of tents made of white clothes besides the huts made of grass.[50] In Harshacharita, Bana presents a vivid picture of royal camp of Harsha. The gates, leading to his palace while in camp, were evidently thronged with elephant and camel corps. These elephants were decorated with banners, clothes, kettle drums, conches. In other parts of the royal camp were troops of camel either sent as presents or brought back in return for others. Another part of the camp looked all white with masses of white umbrellas. In another side were conquered and hostile vassal chiefs. The royal camp was, infact, an extraordinary cosmopolitan gathering. It contained not only notable kings of various countries but also Jains, brahmanas, students, foreigners and ambassadors from alien lands.[51] The subjects and king with all their retinues joined it. The camp comprised of four different courts. In the fourth court, the king gave audience, under a pavilion seated on the throne, a pearl like stone, with his feet resting on a foot tool of sapphire and ruby. The entrance of the gate was regulated by the pratihara and the chief of whom was called pariyatra. Inside the camp was a stable filled with king's favourite horses as well as the state elephant named Darpasata.[52]

Principle of Morality

With the expansion of empire and emphasis on new methods of war one finds a change in the conduct of warfare. Earlier, the concept of dharmayudha or dharmvijya as visualized by Manu in the form of warrior fighting with his equal was losing its credibility and other forms of warfare gained preeminence. Even Kautilya who propagated dharmayudha and if not successful then one should take resort to other methods of warfare, whether the method was fair or foul. Later on, the method adopted by the kings were somehow to win over his enemies. Bana states that Sasanka of Bengal, acting as an ally of Malwa king, enticed Rajyavardana, King of Thansewar, by fair promise to a conference and assassinated him when the latter was off his guard.[53] In the 10th century AD Betaza II of western Ganga dynasty is stated to have resorted to similar method of treachery to get rid of the Cola king. Similarly, the women held important position in society. The violation of the rules of ethics were also seen in this case. Bana mention, that the king of Malwa after defeating and slaying Grahvarman, the Maukhari king, cruelly misused queen Rajyasri, and confined her like a brigand's wife with a pair of iron fetters kissing her feet.[54] The Gauda-Gauda, proudly records that Yasovarman killed the king of Gauda, carried the ladies of his harem into slavery and made them ply the chamaras over him in the public darbar.[55]

All this shows that with the passage of time, the change in political set up of the state and the kings ambitious designs forced them to obtain result in war through any means. Kamandaka while supporting Kautilya, boldly states that one need not hesitate in the destruction of the enemy even by the unfair and immoral means.

In the conduct of statecraft, Bana's approach was different from Kautilya because of its immoral quality. Bana makes his observation through Sukunasa. The Kautilya creed of state policy was not acceptable by Bana. He observed: "whatever can be expected as proper in the vision of those whose standard for action is the dreadful treaty of Kautilya which contains advice mostly of a very cruel nature; whose spiritual preceptors are the family priests with nature made pitiless on account of their practice of malevolent magical rites, whose teachers are ministers expert in deceiving others, whose heart is set on Lakshmi, enjoyed and abandoned by thousand of kings, who are skilled in the sastras which instruct the infliction of death and to whom their brothers affectionate by natural feeling are the persons fit to be slaughtered.[56]

The above observation does not go well with the Kautilyan approach to politics. In Arthasastra, Kautilya mentions several ways and methods to obtain political goals. Similarly, Dandin in Daskumarcharita differs from Kautilya. It is well know that Kautilya was the foremost strategist of his time and hence his principles cannot be totally rejected. One should be clear that politics is not the work of sages who are detached from worldly affairs. Hence, the problems of foreign policy cannot be solved only by following the ethical code of war which Bana's immensely supported. One cannot condemn treachery, which king subscribes in the attainment of political and military interests. Even in Mahabharata, it is stated that "in response to the demands of time and place what is proper may become improper and what is improper may become proper.[57] The nature of war had drastically changed and it could not be confined to the rules of ethics of war.

Thus, the king can indulge in such activities if he wants to subjugate the rest of the king's circle under his domination. In the end, Sukanasa mentions that though the king's rule indulges in evil activities and who is not a victim of infatuation caused by youth and owner should endeavour not to be the target of blame. His activities should be such that he could be appreciated by gurus, friends and wise man. He should be well aware of the dangers. By exhibiting his prowess a king should strive to crown himself with glory because a glorious king has his commands fulfilled as the prophecy of a sage who is powerful enough to see the three worlds.[58]

Relationship Between Foreign Policy and War

There is an intimate relationship between foreign policy and war. Focus on foreign policy was to improve of relationship between or amongst the states. Such relations could be possible by using the method of negotiation, persuasion and conciliation for promoting the common interests of different states and adjusting those interests which are opposed. In attaining the foreign policy objectives the diplomacy is frequently used. Behind diplomacy, one states prepares for war and in certain cases, the threat of war is used as method of diplomacy.

Bana, too has thrown some light on the conduct of kings so far foreign policy and war are concerned. This is well stated in the Kadambari which describe the expedition of Candrapida's desire of conquest. It is mentioned that the kings who are reckless, rash and despotic the need is to suppress the enemy. The modest and humble kings are to be protected. The vicious and despotic kings should be extirpated while the timid kings are to be consoled. A tyranny king is disliked by its own subjects and also by other states. Thus, Candrapida was well conversant with all those rules of conduct. In his expedition, Candrapida subjugated all those kings who were haughty and proud on account of adequate resources. He raised the status of humble kings, consoled the fearful and took aggressive action against the agitated enemy.

It is interesting to note that Bana's observation of the relationship between the feudatory kings and the conqueror. He had mentioned two kinds of feudatory kings: (i) those who were defeated by Harsha in the battlefield, and (ii) the kings who themselves approached Harsha drawn by his irresistible glory. Bana has focused more on the defeated kings. He has given a realistic description of it. In Kadambari, Bana mentions that the defeated and dethroned kings were again installed by the victorious king on the condition that they would pay tributes to the emperor. These kings enjoyed almost the same prestige and dignity as before. But certain privileges of these kings were curtailed.

Subjugating the kings, the emperor with prosperous treasury may effectively discharge his duty as an overlord of the feudal states. Not only this, he also wins the affection of the subjects, exhibited his valour, increased his strength, fame and virtues and this shows the good conduct of the king towards his feudual states.

Bana in his Harshacharita gives another picture of the same situation. One finds that the kings defeated in battles were in a miserable condition which stood in the courtyard and waited for a vision of the emperor. These kings were eager to know the programme of the king through king's servants and even observed the status of a captivated slave. As such, the defeated king's territories were annexed by the victorious king to his own.

Thus, in matter of war, this was an usual conduct of foreign policy during expedition of conquest. The same kind of practice was seen in the writing of Kalidas's Raghuvamasa where emphasis was placed on Raghu's digvijaya, i.e., conquest of quarters.

Conclusion

Basically Bana's was a refined and romantic poet. But he was equally aware of the social and political conditions of his time. The change was seen in the political and military systems during his time. The emphasis was on "would be conqueror" or "universal conquest". Universal conquest was based on the development of military power. The reality of his time was that changes were taking place in the army organization. The fourfold army divisions became three fold divisions because of the decline of the chariots. Bana does not mention about the role of chariots. The elephants were given more prominent place as depicted by his writings. The reason, perhaps, was that elephants were regarded as a symbol of status and prestige of the king and also produced psychological impact on the enemy. Later on, the use of elephants also declined because of the development of more advanced and sophisticated weapons. Elephants became a source of weakness as they destroyed their own armies. But, the still, the use of elephants were also seen in the medieval period also.

Another, important contribution of Banabhatta was that he was a practical man who was aware with the power politics of his time. During Harsha's rule, Bana's observation of the relationship between the king and the subjects, integrity of administrative system, caution against retaliation, proper time to attack, generalship and battles are still valid today. Today, we talk of the civil-military relations in conducting the domestic and foreign policies. Earlier also, as in case of Bana, there was also visualization of this kind of relations. The only difference was the nature of political set up. Earlier, everything was dependent on the king's conduct of state affairs and today we have modern governments in the form of democracy, aristocracy or monarchy. Thus, one can say that being a poet, Bana's contribution to military affairs, the king role in the smooth functioning of the state cannot be forgotten and later on, many poets came out with the similar ideas on statecraft.

REFERENCES

1. U.N. Ghosal, *Studies in Indian History and Culture,* 1957, Calcutta, p. 49
2. B. Sharma, *Harsha and his Times,* Susma Prakshan, Varanasi, 1970, p. 328.
3. E.B. Cowell, and F.W. Thomas, (trans), *Harshacharita,* London, 1897, reprint in 1961, p. 36.
4. *Harshacharita,* Chapter IV, p. 352.
5. T. Walters, On Yuang Chuang's Travel in India, Vol. 2, London, 1904-05, 1st edition, 1961, New Delhi, p. 335.
6. B. Sharma, op.cit, p. 82.
7. E.B. Cowell and F.W. Thomas op.cit, p. 82.
8. See Banabhatta Kadambari of Bana, London, 2014, p. 15.
9. Ibid., p. 203.
10. Ibid., p. 334-335.
11. Harshacharita, Ch. II V. 3.
12. Ibid., pp.140-45.
13. Kalidas, Abhiyananasakutalam, Act V, p. 7.
14. Harshacharita, op.cit, pp. 495-96.
15. Mahabharata, V, pp. 37-38.
16. Manusmriti, VII, pp. 30-31.
17. Kadambari, Ek Sanskritik Ashyayan, Patna, 1962, pp. 19-20.
18. Ibid., pp. 200-203.
19. Matysa Purana, p. 175.
20. G. Upadhyaya, Political Thought in Sanskrit Kavya, Chaukhambha Orientalalia, Varanasi, 1979, pp. 119-20.
21. *Kadambari,* op.cit, pp. 364-78.
22. *Harshcharita,* Ch VI, p. 578.
23. *Harshcharita,* p. 606.
24. Kautilya, *Arthasatra,* VII, p. 224.

25. *Harshcharita*, VI, p. 132.
26. Kamandaka Nitisara, XV, p. 51.
27. P.C. Chakravati, *Art of War in Ancient India*, Dacca, 1941, p. 21.
28. *Panchatantra*, p. 174
29. P.C. Chakravati, op.cit, p. 26.
30. P.V. Kane, *History of Dharmasastra*, Poona, 1946, III p. 206.
31. *Harshcharita*, Ch II, p. 185.
32. *Harshcharita*, p. 68.
33. Ibid.
34. E.B. Cowell and F.W. Thomas, op.cit, p. 180.
35. Ibid., pp. 180-82.
36. Ibid., p. 77.
37. Ibid., p 189.
38. Harshcharita, p. 196.
39. Ibid., pp 152, 196 and 213.
40. Ibid., p. 65 and 211.
41. Ibid., p 188.
42. Ibid., p. 209.
43. V.S. Agarwal, Harshcharita, EK Sanskritik Adhyayan, Patna, 1962, p. 40.
44. Ibid., *Kadambari*, Ek Sanskritik Ashyayan, Patna, 1962, p. 32.
45. *Harshcharita*, p. 197.
46. I. Elliot, pp. 140-45.
47. Kamandaka, op.cit, XIX. 2.
48. C.M. Ridding, Kadambari, p. 86.
49. *Harshcharita*, p. 203.
50. V.S. Agarwala, op.cit, p. 268.
51. *Harshcharita*, pp. 46-49.
52. Ibid., p. 50.
53. Harshcharita, Vol. VI, p. 178.
54. Ibid., p. 153.
55. P.C. Chakravarti, op.cit, p. 186.
56. *Kadambari*, p. 383.
57. Mahabharata Shanti, 79/31.
58. *Kadambari*, p. 394.

8

SOMADEVA SURI

Introduction

It was seen that the beginning from 8th century onward war was generally regarded as a period of intermittent and internecine struggles. The numerous powerful Rajput kings tried to established the suzerainty in various parts of northern India. Such supremacy was possible by ample demonstration of chivalry shown by the kings and soldiers and better preparation of weapon systems. Forts, too, played an important role in securing the king and his soldiers. One also saw the newly emerged kingdom's territories were not defined and emphasis was on the expansion of their territories and the acquisition of enemy's wealth. More or less, the kings felt in secure from their neighbours. Thus, the prime responsibility of the king was the protection of the kingdom from external and internal threats. For the defence of the kingdom, army played a pivotal role. Almost all writers of this period emphasised the need of army, though army was regarded by them as one of the essential elements of the state. As such, the political and military writings of Somadeva Suri hold special position among his contemporaries and even today.

Somadeva Suri was a Jain scholar. He wrote the famous 'Nitvakyamritra' or the nectar of propositions concerning niti. His work is written in sutras (aphoristic style). It is believed that the work was probably composed in 1006 AD. It is a political work. It is divided into 32 chapters and it contains 1525 sutras. His work had the impact of

Kautilyan and Manu's writings on statecraft. U.N. Ghosal very succinctly remarks, in matter and in form it agrees most closely with Kautilya's Arthasastra.[1] Beni Prasad also agrees that Somadeva Suri Nitivakyamitra is heavily indebted to the same old writer, Kautilya.[2] As an exponent of statecraft, Somadeva Suri follows the Hindu tradition. In an age of political decadent, he cries back to Kautilya. He often borrows the thoughts and sometimes the very expression of his master. Frequently, he refers to Manu, Bhimsa, Bharadwaja, Visalaksa, and other writers. But his mastery of a literary craftsman enables him to weave all his various collections into a fine and harmonious whole, which has all the appearance of unity and originality.[3]

Though Smritis and Mahabharata can be classified as pure political work but it is only Kautilya's Arthasastra that represents the entire ancient niti literature. So far Somadeva's Nitivakyamrita is concerned it well connects with Kautilya's Arthasastra and Kamandaka's Nitisara. It is also true that after Kamandaka Somadeva wrote pure political work. The literary expressive thought is well conceived in the theory of statecraft. As an exponent of statecraft, he has discussed that art of war in detail during peace and wartimes. Moreover, his work is a unique combination of sociology and statecraft. Thus, the theme of state, society, relationship between the king and state and inter-state relations have been fully developed by him because the age in which he propounded his views was that a balanced and secured relations must exist so that the state can grow and proper. Understanding these problems, he took an independent approach and laid emphasis on basic values of philosophy. Dr. M.L. Sharma, on the basis of ancient statecraft attaches first place to Kautilya, second to Somadeva Suri and Kamandaka to third place.[4]

Somadeva discusses the importance of dharma, artha, kama, six kinds of passions, acquisition of knowledge, four types of knowledge (Rajavidya)—dialectical philosophy, vedic theology, agricultural economics and the art of punishment—ministry, priest, commander-in-chief, envoy, the spies, policy deliberation, vyasanas, king, amatyas, janpada, forts, treasury, army, friends, protection of the king, civil law and conflicts, the saptang policy and the method of fighting so that stability and prosperity can prevail in the kingdom. Prudence, realism and moderation appear to be the guiding precepts of his works. He follows the method of analysis of experience. Hence, his work contains main maxim, which can serve as basis of philosophy of contended life which is important for growing state.

Three Goals of Human Life

While explaining the importance of dharma, artha and kama, Somadeva states that the problem is people centric. To him, dharma and artha are equal in importance. He also equates kama with dharma and by doing so he has showed his political maturity. Being a Jain, he experienced the significance of niti. In Nitivakyamitra, he explains dharma as the means for the realization both of prosperity (abhyudaya) and spiritual good (moksha). Sacrifice and religious austerity are the means that leads to dharma.[5] About artha, he writes that a man should fulfil all the needs of life.[6] Moral beings who try to obtain unpossessed wealth and try to defend his wealth and increase it would make his life happy and prosperous. The utility of wealth was also well taken by earlier ancient thinkers. They believed that lusty people who do not donate his wealth would lead to destruction. Somadeva mentions people who sacrifice his life for their own comforts should not accumulate wealth because it is against niti. The real utility of wealth is possible only when it controls man's mind and senses.[7] About kama, Somadeva cautions the king to do rightful duties. A king who fails to overcome kama cannot protect the various elements of state against his opponent. Thus, a moral man should work properly according to dharma, artha and kama. Giving importance to any one would lead to self pain.

Thus, the purpose of a man is to perform the welfare of the three goals of human life (dharma, artha and kama). The welfare of human being by the king is possible by the adoption of the policy of punishment (danda). Somadeva says danda should be used in a just manner. Irrational punishment would make the people unhappy. If a king does not make use of danda a powerful king would not crush the weak. By pursuing danda not only the state and society are protected but also the subjects can attain the three goals of human life.

State

Unlike Somadeva, Kautilya and Kamandaka do not define the term state. They have mentioned the various elements of states. Somadeva, while describing the various elements of state, also defines state. In one place, he writes that the king's rightful duty of the state is the protection of the land.[8] (prithivi). Further, he says that the state is comprised of brahmins, ksatriya, vaishas, sudras and ashrams and land where grains, animals, metal, iron etc are found in plenty is called a state.[9]

The above definition covers the broad aspect of state. In the first definition, there are three main elements of state, viz., the king, land and the duty of the king to protect the land. According to him, the main ingredient of state is land. But this land should be a cultivable and prosperous land. Uncultivable land and without population land cannot be called a state.

The above assumption leads to an important question, which work is capable of protecting the land and who can do so? This is amply demonstrated that the land is protected by army and wealth. Thus, for the protection of land, warriors and patriotic soldiers have to be organized into a coordinated body. The responsibility of protecting the land is the work of the king. The salary of the soldiers and maintenance of ammunition needs wealth (kosa). Not only the army but the entire government machinery depend upon wealth. Thus, for the defence and prosperity of a state wealth is necessary. The main duty of the king is to maintain good agriculture and through it develops prosperous wealth. Somadeva mentions kosa, danda and bala are the powers of a king.[10] Besides these elements, Somadeva mentions the importance of amatyas because the king alone cannot perform the necessary functions of the state.

In the second definition of state, Somadeva lays special emphasis on population. Without population, state is not a state. People are important than animals and birds because it is the people which forms the community. The most important question is how much population a state should have? But one thing is clear that more populated state and rich in natural resources will make a state strong. Somadeva does not mention number but still, population is important. Thus, kosa and population have intimate relationship. Lack of kosa and population will lead to instability in the state. To him, kosa is the soul of state and lack of it no state can function properly. In this way, second definition appears to be quite scientific and has added utility.

Origin of the State

Western writers have propounded the following theories of the origin of state. These are: (i) divine theory; (ii) force theory; (iii) social contract theory; and (iv) historical or developmental theory. The Indian thinkers have laid emphasis not on the theory of the origin of state but on the origin of the theory of kingship. In ancient India, the idea of kingship emerged due to chaos and confusion among the subjects and the duty of the king was to remove these so that the society can perform its function properly. Thus, in ancient Indian literature, generally the theories of kingship is as follows. First, the divine theory where the king is responsible to god. It is well established in Mahabharata and other Smritis. Second, the social contract theory is found in Bodh literature and Arthasatra where the emphasis is on the consensus among the people. The third is the vedic theory who believed that the origin of kingship was due to the necessity of a leader in war.

Functions of State

Somadeva has mentioned the functions of state in Nitivakyamitra. Modern theorists have divided the functions of state into two parts: (i) essential, and (ii) non-essential functions. Essential functions revolves around the state's security and the relationship between the states. The duty of a king is to protect the state from external and internal threats otherwise it will lead to destruction. The performance of king's duties is depended upon a well-established army, police force, appointment of state employees and proper functioning of judiciary. Non-essential functions include education, public health, agriculture and trade, transportation system and economic security.

Somadeva has mentioned both functions in his work. According to him, protection of good people and destruction of evils from the society should be the main duty of a state.[11] He writes those king who do not show valour against the enemies are contemptible.[12] The duty of a king is to punish the guilty according to crimes. Lack of danda leads to the existence of Matysnaya. Danda compels the people to perform their tasks properly. Besides mentioning the essential functions Somadeva also lays emphasis on non-essential functions. Ignorance of them will not lead to the proper development of the state. Thus, the policy of a state should be the welfare of the people. In other words, the promotion of agriculture, trade, etc., become the responsibility of the state.[13] Thus, Somadeva has emphasised the functions of state in terms of material and spiritual progress. The modern writers too have spoken of a welfare state.

Elements of State

Modern political and strategic theorists have mentioned four basic elements of state. R.G. Gettell has mentioned population, territory, government or administration and universal state as the main elements of state.[14] A state, then, could be defined as a community of people, permanently occupying a definite territory, legally independent of external control and possessing an organised government which creates and administers law over people and groups within its jurisdiction.[15]

In ancient India, scholars have mentioned seven elements of state. All these elements constitute a state. These elements are king, amatya, janpada, fort, kosa, danda and friend. They have compared seven elements with the organs of human body. Thus, these thinkers developed savayav (composed of parts) form of state. In the western world it was developed in the 19th century. Prof. Bhandakar,[16] Dr. K.P. Jayaswal,[17] Prof. B.K. Sarkar[18] and others believed that ancient thinkers made a systematic study of savayav form of state. Such studies also analyse the importance of the theory of balance of power. It meant a friendly relationship amongst the states so that a friendly environment could be created. It would be appropriate here to discuss these elements of state in detail.

King

King is important because people want a secured environment where their problems are solved and the dignity of the king is also maintained. Almost all ancient thinkers tried to equate him with gods. Kalidas in Raghuvamsham has described Raghu as a king whose presence has even made moon significant and made people to carry out his religious austerity and has also made sun purposeful. Similarly, Raghu while making people happy made the king more important.[19]

The importance of a king lies in the performance of his duties. He is an exponent of dharma, artha and kama. The main dharma of a king is the elimination of evil forces and protection of good people. The function also includes the protection of the people from external and internal threats. In ancient literature, it is mentioned about the necessary requirements of a king. It is said, when there was no king and a state of chaos and confusion prevailed in the state then Brahma, in such circumstances, created a king for the protection of the universe.[20] Similar view is echoed by Kautilya and Kamandaka.

The importance of a king also lies in evolving the natural or the materialistic world. Lack of materialistic world will not get the desired result. Somadeva illustrates this by saying the losses incurred by the state without a king is similar to a tree whose roots have been uprooted and expects that tree will blossom but this will not happen.[21] In such circumstances both the king and the state would be destroyed.

Origin of Kingship: In ancient India, various theories emerged regarding the origin of kingships. These are: (i) vedic theory; (ii) social control theory; and (iii) divine right of kingship.

Vedic Theory: This theory is the oldest theory. According to it, the origin of the king arises due to the necessity of a leader which could lead an army in battle. It has been discussed in detail in Aitraya Brahmana.[22] There is a reference of a battle between goods and devils. While losing the gods analysed the causes of it and found that the devils were successful because they had a king who could lead them in battle. Thus, later on, Aryans decided to choose a person who could be the first king.

Social Contract Theory: This theory revolves around the development of consensus among the people. This theory is found in Bodh literature and Kautlya's Arthasastra. The theory became popular because of a state of confusion and political chaos in society. The people, then, realised the necessity of an able person as a king who could protect the people and bring prosperity and security in the state by punishing the offenders. In modern times, Hobbes, Locke and Rousseau propounded this theory by emphasizing the people's participation in choosing the government.

In Kautilya's Arthasastra, it is mentioned when the subjects were afraid of

Matsyanaya they, then, appointed Manu as their king. Once chosen, Manu took the responsibility of the welfare and defence of their subjects.

Divine Right of Kingship: The theory owes much to god. The theory is well stated in Dharmsastra and Puranas. In Mahabharata, god and king are equal.[23] The king is supreme and is responsible to god and not to the people. But the king's responsibility's lies in securing the life of his people and in this way people does not also disobey the orders of the king. In Dharmsastra, it is mentioned that the king is the creation of various particles of god. Manusmriti states that god inorder to protect the world created a king and equated him with the powers of Indra, Vayu, Yam, Surya, Agni, Varun, Chandra and Kubera.

As such this theory flourished till the Gupta age. Somadeva believed in this traditional theory. According to him, king was the symbol of Brahma, Vishnu and Mahesh. Thus, there were no other god.[24] Accepting Bramchar and following all norms of gurukul, such a kshatriya king is like Brahma. Acquiring the qualities of Lakshmi and working for the welfare of the people such a king is like Vishnu. Fighting and destroying the demons such a Vijigisu king is equal to Mahesh. Thus, the king, acquiring all the qualities of three gods, have found its supreme expression in Nitvakyamitra.

Qualities of King

Ancient Indian thinkers have discussed in detail about the good and bad qualities of a king. These qualities are that the king should control over his senses, be a great warrior, exponent of nitisastra, good understanding of triad as laid down in the three vedas, anvikshiki, varta and dandaniti courageous, religious, simple, obedient, skilled in sama, dama bhed, and danda and application of six-fold policy.

Knowledge of three worlds enable a king to permit his subject to follow their own religion independently. By doing so, both the king and the people attain dharma, artha and kama. Somadeva mentions that a state can prosper while promoting agriculture and trade. Like Yamraj, a king should take serious measures in punishing the offenders. A king should show his skill in conducting the performance of the state.

Nityakyamitra mentions about a just king. A just king will always make the people happy and prosperous. He is also capable of resisting any revolt and bringing prosperity and security in the state. For the proper functioning of the state knowledge of sama, dana, bheda and danda are necessary. A king unknown to these rules will not succeed in expanding his dominion.[25]

Somadeva also mentions bad qualities which a king should avoid. There are anger, arrogance, lust, ego, wickedness, foolishness, unrighteousness and cowardice. A king endowed with these qualities is like a mad elephant and is not worthy of a being a king. Foolish king has been criticized by him. He states a state without a king can be imagined but a foolish king is not good for state's prosperity.[26] A king endowed with vices can also harm the state. Here Somadeva mentions 18 vices and a king possessing any one of them, though he has chaturangabala, is destroyed. An arbitrary king who promotes his own self-interest cannot bring happiness and prosperity. Such a state cannot be secured for a very long time. While explaining pride, Somadeva says it increases valour.[27]

Thus, Somadeva took full liberty in pointing out the good and the bad qualities of a king. These qualities of the king spell out the advantages and disadvantages of the growth of the state. Moreover, he says a good king can attain divinity.

Duties of a King

A king is born to remove anarchy from the society. The responsibility of the king lies in providing peace and security in the state. Thus, in place of state of anarchy peace has been given importance. This is the prime duty of the king. The other duties include: (i) protection of subject from external and internal threats; (ii) establishment of social system; (iii) economic prosperity; (iv) administrative functions; and (v) to deliver justice. Moreover, ancient thinkers have emphasised a state is protected by dutiful king and his subjects. In this way, the protection of the king becomes important because the protection of the subjects depends upon the protection of the king. Here, Somadeva, too, subscribes this view.

Ministers

Somadeva believes a man of its own cannot perform the function properly. Thus, it is essential that the king needs large number of persons to perform his function dutifully. The king, before starting any work, should hold consultations with his ministers. Since one or two persons are capable of rendering goods advice to the king, then he can appoint 3, 5 or 7 ministers who can impart knowledge on different subjects. Like in Sukraniti, Nitvakyamitra also stresses the quality of ability regardless of caste. A minister should possess the following qualities; birth in the state, righteous conduct, good family, freedom from vices, knowledge of arms, good behaviour and purity from the Upadhas of dharma, artha, kama and fear (dharmopdha, arthopadha, kamopadha and bhayopadha).[28]

Thus, Somadeva has given enough importance to ministers and amatyas. In Nityakyamitra, both the terms have been used. Somadeva makes some distinction between ministers and amatyas. There is a possibility that he has used minister as prime minister and member of inner council and amatyas has been referred as high state officials of mantriparished. He defines amatyas as one which has been given the responsibility to assist the king in fulfilling the purpose of donation and honour.[29] Officers assisting the king in the performance of state's welfare are called amatyas by Somadeva. This distinction is also found in Kamandaka, Agnipuran and Kautilya's works.

One thing is clear that amatya, being a member of mantriparishad, does not possess the right of deliberation. This right is possessed by trained and credible ministers. Though the member of mantriparishad was large but inner council has only 3 or 4 ministers. Somadeva states the combined effect of the separate abilities, a large number of ministers increase the efficiency of the government. However, when one minister possesses all the faculties necessary for the proper administration of the kingdom, there is no harm in having only one or two ministers.

According to him, the scope of consultation is as follows: knowing the unknown subject, decision on the known matter, to resolve upon the decision, not to leave any doubt in regards to points of differences etc.[30] Somadeva holds that consultation consists of five elements, which are: (i) means of starting works; (ii) assembly of capable persons and materials for them; (iii) doing a work with the consideration of country and time; (iv) protection against obstacles in the way and (v) achievement of the desired objective. He has also emphasized the importance of maintaining secrecy in relation to consultation.[31]

Janpada

From political perspective, the term janpada has been used in the context of organized community. In Bodh literature there is reference of 16 mahajanpadas. Panini and Kautilya also mention janpada. To them, protection of the janpada means defending every village. In the frontiers janpada, forts are to be constructed. Somadeva has discussed in detail about janpada. Janpada is best when it protects the king and vice-versa. Janpada that is full of gold, silver, metal, iron, sulphuric acid and elephants, where the population is neither huge nor small, where good materials and different kinds of grains are found in plenty and for trader the importance of export-import qualities makes people and animals happy.[32]

People live in those janpada where the basic necessities of life are easily met and hence they are secured. But, on the contrary, lack of these qualities prove harmful to the king and his subject. People then, settle in other parts of janpada. According to Somadeva, that janpada is not worth living where the means of living of people are not available. Thus, rational people should settle in those places where the essential necessities are met.[33] These qualities indicate the growth of janpada where people can find happiness.

Somadeva also points out the demerits of janpada. These demerits are uncultivable land, rocky and hilly areas, full of caves, abode of snakes, bhil and barbarian, and growth of small quality of grains compelling people to live by eating fruits.

Somadeva also mentions about the settlement of janpada. Those people settled in foreign states and who are taxed should be given every benefit so that they can come and reside in his kingdom. In every janpada, village is the smallest unit. Proper care must be taken to settle the people in proportion of different castes in villages. Janpada which loses this balance and gives priority to one particular class become the cause of friction and, thus, leads to uprising. It is proved that in such villages which are dominated by ksatriyas and are more in number lead a miserable life because they fight amongst themselves.[34]

Somadeva states a king should not give that village to anybody which is a cultivable land. Such villages provide means for the fourfold army of the king.[35] If the king gives the village the army will not get proper supplies of grain. Without it a king will not be able to maintain a huge and powerful army. Lack of proper army will put the king in a difficult situation as he will not be able to defend the kingdom.

Forts

Ancient Indian history is full of significant battles. Battles were important for the expansion of a king's state territories. Even the Chakravarti king fulfil their desire through universal conquest. Thus, the strength of the kingdom lies on the construction of various types of forts suitable to the king. The forts were constructed in accordance with the suitability of the topography of the king. Thus, ancient writers had visualized fort as one of the important elements of the state. A king having more forts was regarded as a powerful king. Kautilya and Sukracharya have discussed in detail about the utility of forts. Similarly, Somadeva realizes the significance of these forts. Fearing the enemy's attack the king constructs forts, trenches, etc. By doing this, the enemy is placed in an indifferent position because the enemy's attack can be contained.

While emphasising the importance of forts, Somadeva states a kingdom without forts amounts to defeat. A fortless king has no one to defend him just like an isolated bird in the middle of the sea has no place to settle. Kautilya writes if there is no fort the enemy can easily acquire kosa. Thus, the construction of forts becomes important to defeat the enemy and to save the kosa. Even, army can easily be secured from these forts. Those kings possessing strong forts are well protected and difficult to defeat them.[36]

Somadeva has classified forts into natural and artificial forts. Natural forts are of four kinds – Audak, Parvata, Danardurg and Vandurg. Earlier Manu has mentioned six types of forts, Sukracharya, seven types and Kautilya four types.

1. Audak forts: These forts are surrounded by all sides from river and in the middle there is an island which is situated or surrounded by large lakes.
2. Parvata forts: These forts are made of large rocks or situated in such difficult places in the form of caves.
3. Danavdurg: These forts are situated in difficult places and where grass, water, landless or uncultivable land are found in plenty.

4. Vandurg: These forts are surrounded by mud or barbed fences.

The purpose of water and hill forts are to protect the kingdom from the enemy's attack. Dhanv forest durgs protects atvika. In a crisis, the king used to take shelter in these forts.

What are the qualities of forts. Forts help the king in crushing the uprising of the enemy and thus protects its own kingdom. He explains it in this way that because of hills become difficult for the army to move in even and uneven and wide areas. Those forts are good where grass, fuel and water are found in maximum quantity but the enemy is deprived of these provisions. Another quality of forts is that it should be surrounded by brave warriors. Where there is lack of these provisions that fort is like a prison cell which can prove fatal to the king.[37]

Somadeva has thrown light on the acquisition of enemy's forts by various methods by the king.

1. Abhigaman: It means adoption of conciliatory measure while entering the enemy's fort with his own well-equipped army.
2. Upjaya: By various measures a king should create differences in enemy's amatyas and other officers.
3. Chirnirband: This means surrounding the enemy's forts by his own forces for a very long time.
4. Avaskand: This is done by giving the enemy plenty of wealth and respect.
5. Tikshnapurushaprayoga: This is done by sending deadly spies to enemy's king.

These are five measure mentioned by Somadeva in the acquisition of enemy's forts. The ancient writers too believed in these measures. Equally important in the destruction of enemy fort is to sow dissension within the enemy camp.

Somadeva has given an interesting account in entering the forts. He states that a king should not allow persons to enter the fort who does not have state currency and who are completely unchecked. No such person is also permitted to leave the fort.[38] He has presented some historical fact to prove it. The king of Huna kingdom send his soldiers as traders to enter the fort and by then murdered the king and occupied the kingdom of Chitrakut. Further, Somadeva cites that the enemy king sends his well trained soldiers on the pretext of hunting for serving the king of Kanchi while in the process they enter the fort, killed the king Bhadra and subjugated the kingdom of Kanchi.

From the above discussion it is clear how the forts played an important role in the protection of the kingdom and its king. The construction of forts were not only confined to ancient period but we also saw their use by the Rajput, Mughal and Shivaji. The importance of forts declined when the Britishers entered India with better offensive arms. Thus, these forts became vulnerable and lost its utility during war times.

Kosa

Kosa also constitutes one of the most important elements of state. Somadeva writes that kosa is the soul of the king.[39] The king is capable of protecting his kingdom because he has huge kosa. Kosa also enables the king to engage the enemy for a long time. A king stationed in fort and then fighting a defensive war kosa becomes the source of king's strength.

Kosa is accumulation of wealth by king during peace and war times. By doing so the efficiency of the functioning of the governmental machinery is increased.[40]

Somadeva mentions those kings who do not increases its wealth do not have a bright future.[41] Even Kautilya has laid emphasis on the acquisition and protection of wealth. Similar voices are echoed in Mahabharata, Kamandaka and others. Somadeva states that kosa nor the physique of the king is the cause of his progress. Moreover, a king is victorious because he has wealth. By kosa, army can be maintained and without it a state cannot be protected. All ancient scholars agree that for the prestige, for defence and progress of the state, kosa is very important. Somadeva writes that for a kingdom the best kosa is one where gold, silver and money are found in plenty and in time of crisis capable of spending large amount.

Economically, Somadeva mentions of those qualities of kosa which are important for the maintenance of state. Selling things easily and selling small items at great cost should be encouraged and kept in rajkosa in great qualities. Money becomes important for the payment of the salaries of army officers and other officials. If enemy's king attacks a kingdom and he does not possess sufficient army or feels defeated in the battle then a king with sama and dama return wealth to the enemy. An enemy could be satisfied only when the king's kosa is full of gold and silver. On the other hand, he criticizes those kings who are without wealth. He says a woman leaves a poor man, then, what to say of others. For simple poor men without wealth nobody come to serve him because there is no hope of living. On the other hand, a characterless man having wealth all the people are ready to serve him. Somadeva gives the examples of a pond. There is no point of its expansion where there is dearth of water.[42] But a small pond full of water is more appreciable.

On the question of income and spending, Somadeva states a king should spend money according to his needs because if the king spent more than he earns his position is similar to the life of beggar. According to him, a person spending the money in equal ratio finds himself in a state of happiness. According to Sukracharya, a king should spend 1/6 part of his annual income on army, 1/12 part on donation, ministers and other state officials and his own personal works. It means that a king should not spend more and use the money carefully. The question is how to increase wealth? One method is to levy taxes on the subjects. But taxes should be imposed on the people in proportionate ratio. Somadeva believes that increasing taxes will weaken the state and thus the state will be destroyed. A just king has to give proper thought on this issue. The wealth of the subject is the wealth of the king and hence it has to be used very carefully. Taxing the people too much is not good for the king and the state. Cutting a tree will give fruit only once but not in the future.[43] Thus, taxes should be imposed according to the needs of the state.

It is also the work of the king to work for the welfare of the people. He should be tolerant otherwise intolerance will ruin the state. The acquisition of wealth is important because in times of crisis the life of the people and army are depended. Thus, in Nitvakyamitra it is mentioned that strengthening the economic position of the state, a king should be aware of such situation. He says that the basic needs of a state are agriculture, trade and animals. His thought is even relevant today.

Army (Bala)

Army constitutes another important element of state. Without it, a state cannot survive. The army was variously termed as sena, pritna, vahini and parthiva sainya (royal army). Towards the latter part of the 12 century AD astra vidya and yuddhatatva (military science) formed an important course of study for the kshstriyas. Army is capable of defending the state's territories from external and internal threats. Somadeva analyses the role of army during wartime. To him, it is better to avoid mutual hostilities by peaceful method. Even of the two, army and kosa, Kautilya favoured kosa. Army", he says, "may go to the enemy or murder the king and bring about every kind of trouble but kosa is the chief means of observing virtuous acts and of enjoying desire."[44] War is a utilitarian business meant only for specific results. Where it is possible to buy peace, a king should do so. War should be engaged in when the king is in a far superior position to his enemy. Otherwise, victory in battle being dubious, it is wiser to avoid war. Kamandaka is very clear that the use of organised violence can only be the last resort for any king. Mostly the king should confine himself to sama and dama and if necessary bheda. Somadeva, too, supports Kamandaka and see war as avoidable at best.

But, war or battle becomes a necessity. History is full of examples where state expanded their territories by waging battles. In such circumstances, a king with complete preparedness must destroy the enemy to protect his subjects and territories. Ancient scholars have regarded as good to die while fighting. According to Manu, for the protection of the subject, a king should not run away from the battlefield and if he dies while fighting he will go to heaven.[45] Mahabharata too maintains this view by stating that a kshatriya dying at home is a great sin. The same idea is also visualized by Somadeva. He states that a king afraid of enemy's attack should not run away from the battlefield and leaves his kingdom but should lay down his life while fighting.[46]

King in Relations to War

Regarding war, Somadeva has given some directions to a king. A king should avoid anger. Sometimes anger compels a strong enemy to wage a battle. In such a situation his destruction is fixed.[47] When the king knows he cannot escape destruction he should not run away from the battlefield death is imminent.[48] He believes in waging battles to the fullest extent. Running away from the battlefield, victory and defeat, life and death are subordinate to destiny.[49] He further says that if the enemy is more powerful then a king should not indulge in battle. Instead, he should make alliance with him. Just as foot soldiers engaged with elephant force are destroyed similarly a powerless king is also destroyed by more powerful king.[50] During wartime unknown person coming from enemy side should not be included in one's army. If it is necessary to include him he should be thoroughly checked but he should not be allowed to stay.[51] The unhappy relatives of the enemy should be included after careful checking because this is the main formula of destroying the enemy.[52]

Relations with Feudal Kings

In ancient India, the system of feudalism was quite prevalent. The digvijai king having conquered the enemy did not include them in his state but once they accepted his suzerainty these feudal kings were given internal freedom. Under the feudal system both big and small feudal kingdoms existed. But the Vijigisu's relations with them were not the same. But one thing was common that they follow the king's orders, pay annual taxes, assist the king in wartime and present themselves in raj darbar from time to time. The necessary information of these feudal kingdoms are gathered by spies everyday. In Nitivakyamitra though no detailed information is mentioned but in it there are enough proof that the feudal system existed in Somadeva time. There is description of Vijgisu under whom many feudal overlords accepted his suzerainty.

Types of War

Most ancient Indian scholars have classified war as follows: (i) dharmayuddha and (ii) Kutayuddha. Kautilya mentions another type of war, i.e. tusniyuddha. Somadeva mentions only two types of war. While explaining Kutayuddha, he writes that showing his interest in attacking and returning back and then destroying the other enemy is called kutayuddha.[53] Tusniyuddha is that kind of war where persons expert in giving poison, are send there or going quietly to enemy's side and by means of dissension policy destroys the enemy.[54]

War cannot be waged without the proper development of army. Hence, army has been regarded by ancient Indian scholars as one of important elements of state. According to Kautilya, the necessity of army arises when the state suffers from external and internal threats. When these threats are contained successfully the state, then, is completely secured. Somadeva mentions just as root, branches, flowers and fruits help in the growth of a tree, similarly state grows due to the strength of army's valour and chivalry.[55] He further mentions that a king who does not indulge in battle with the enemy is like death person, even he is alive.[56] To destroy the enemy, the king needs a powerful army.

The main purpose of organization of army is not to harm its own subjects but it should defend the territories of its kingdom. Here, what Somadeva wants to clarify that disorganized army can harm the subjects. If this is so, the dissatisfied subject will hate the king and as such the state will be ruined.[57]

Classification of Army

Thearmy has been explained in Nitivakyamitra. The purpose of it is the destruction of the enemy's forces, accumulation of wealth, donation or fulfilment of the needs of the king and assistance in time of crisis is called an army.[58] In ancient time, chaturangbala was much in vogue. It meant the army composed of four forces. These were elephants, horses, chariots and foot soldiers. In chaturangbala much importance was given to elephants.[59] Somadeva believed that main cause of a king's victory in battle was an elephant, because soldiers who fought sitting on the elephants and fired arrows were capable of fighting many soldiers.[60]

Elephants: Nitivakyamitra mentions the importance of elephants in battles. They were important for four reasons: (i) being physically fit and powerful; (ii) chivalrous and if this quality is missing it is no good for waging a battle; (iii) trained elephant and ill-disciplined could be the cause of strength or weakness of own's army and (iv) usefulness of them could be seen while wandering in difficult places and by it the victory is possible.[61] Kautilya too mentions of the functions of the elephants and there is independent department of elephants under Hastadyaksha. The functions of the elephants are: clearing the path, helping the army in movement, entering the water to know the depth of the water, forming a strength line when attacked by enemy's forces, moving from high to low places, moving in jungle and making scathing attack on the enemy, winning, uniting the diversified army, creating chaos in enemy's compared ranks, creating fear among the enemy's soldiers, protecting the kingdom in crisis, helping in catching an enemy's soldiers and freeing its own soldiers from prison cell, showing efficiency in battle and finding a weak point of enemy and killing even the sleeping soldiers. Thus, elephant was the main force while the other three occupied a secondary position.

Cavalry: After elephant, cavalry occupied a dominant position. It would be to underestimate the role of cavalry. Their role was equally significant. Cavalry provided mobility and that to emerge victorious, a king must be strong in cavalry. The Chalukyan Emperor of Somesvara (1127–1138 AD) comments in Manasollasa that strong cavalry force is essential for conducing offensive and defensive operations. While praising cavalry, Somadeva writes that these kings who give prominence to cavalry is like Lakshmi which makes people happy because they get plenty of wealth. Long distance enemy shows closeness to the king. The performance of cavalry is seen in the battle and the king gets the desired results. Cavalry generally surprises the enemy and the enemy forces are scattered all over. Somadeva says this is the work of cavalry and not chariots. Sukracharya also attaches much importance to cavalry because a king along with cavalry attack the enemy front and then enters the long distance enemy's territories.

In Nitvakyamitra, some light has been thrown on various races of horses. Earlier Arthasastra had classified horses in three various types, i.e. Tiksna (fiery), Bhadra (gentle) and Manda (sluggish). Of them, the first two types were used for military purposes. The role of the cavalry included "exploration on occupied positions, camps and forests, holding strategic positions, water fording places and positions destruction or protection of the commissariat, of arriving fresh troops, supervision of the discipline of the army, lengthening the line of the army, protecting the sides of the army, first attack, dispersion, trampling it down, defence, seize, causing the army to take a different direction, carrying the treasury and the princes, falling against the rear of the enemy, chasing the timid, pursuit and concentration.[62] With the passage of time its utility in battles increased immensely. Somadeva says that war had become a sport as the enemy at the long distance could be reached easily.[63]

With the passage of time cavalry was gaining more prominence than elephants and chariots. The cause of Porus and Rajput defeats in various battles were due to elephants because the muslim invaders were equipped with best horses. Anandpal, son of Jaipal at the coast of Sindhu river fought against Mahmud Gaznvi. The Rajputs could have defeated Mahmud if his own elephant had not left the field and that became the cause of their defeats.

Chariots: This was the third important part of chaturagbala. The Mansollasa of Someshvara, a work of the 12 cent AD, continued the application of the term although chariots were not regularly used in warfare from the 7th century onwards. However, the inscriptions of Yashkarana of later Kalchuri dynasty dated February, 1076 AD refers to the traditional four member army; the chief ministers of the lord of the earth placed on the throne his son Kokalla (II) lion like prince, the advance of whose four-member army was checked only by its encountering the multitude of waves of the four oceans.[64] Yuang Chwang, too, speaks of Indian army as consisting of four arms namely, foot, horse, chariots and elephant soldiers.[65]

Somadeva mentions that chariots were useful in plain ground where soldiers laced with bow and arrow were able to defend itself from the enemy. In chariots the role of sarathi was important because entailed giving directions at proper time and place and generally it resulted in victory. While describing the importance of chariots Somadeva states when a soldier skilled in archery fights in plain battlefield and launches attack on the enemy then there is nothing impossible for the king.[66] On the other hand, uneven ground becomes the cause of defeat.

According to Somadeva sarbhut army should be placed in the front. With them victory is possible. Chariots of king are in a position to destroy the enemy more easily. Thus, he places utmost importance to sarbhut army than traditional maula army, allied, sreni and atvika armies. To him, saraheen army is a weak force because there is lack of skill in warfare and it thus becomes the cause of defeat.[67]

Infantry

The last division of army was infantry. Generally, it consisted of four types of soldiers, i.e. maula, bhrtaka, sreni and mitramitra. Another classification was two fold, i.e. anyagama (soldiers coming from others) and suagama (soldiers coming from king's own country). Infantry is at arm best suited for closed combat. In the mountainous region, the infantry with its light equipage and ease of movement rendered good services by effecting diversions, penetrating well defended areas and laying ambushes. The forts were also defended by the footsoldiers from the battlements and parapets. Footsoldiers generally comprised of brave kshatriyas class who were skilled in art of warfare and proved useful in battles.

Military Commander

Somadeva has thrown sufficient light on military leadership. It is the responsibility of the king to appoint a commander. In ancient times a man possessed of distinct military qualities was made a commander. This position was very important. Therefore, everyone was not suited for this post but possessing specific qualities were made a commander because victory and defeat of the king and the state depended on him. Somadeva mentions the qualities of commander as follows: gentle, good conduct, skilled in statecraft, loyalty, purity, skilled in the use of all moral rules, capable of conducting all types of force (elephant, horses etc), powerful and courageous, knowledgeable, thinking of victory, a king who has honoured and by giving wealth made him famous and capable of bearing all kinds of sorrows.[68] A commander possessing all these qualities makes a king great otherwise lack of it will yield no result or defeat in the battlefields.

Somadeva also points out the demerits of the commander. These are: whose nature is one of being unsuccessful in battle, valourless, uncontrolled over his senses, arrogant, expendable, living in foreign soil for a very long time, creating obstacles amongst the people and within his own army, eager to know unnecessary talks, spending salary on his own, critics of king's work, acting against the wishes of soldiers and jealous of the problems of state's interest.[69] Such persons should never be appointed a commander by the king because doing so it will be a cause of great destruction.

Duty of King towards Valorous Army

The behaviour of the king should be good towards his army and state's officials. If not, they will not support the king. A king should not dishonour his army but giving respect make them happy. Along with it enthusiastic (encouraged) army who has attacked the enemy and the other state's army who has invaded his territory should be made happy by wealth and honour.[70]

According to Somadeva, king's maula army support him even in crisis and even on being punished it should not revolt and should not be a part of enemy's army. Thus, a king should satisfy them with wealth and honour.[71]

A soldier gives more importance to honour than wealth. If the king respects his army and is contended with his great work then he is ready to defend the state with great strength. This respect will instil the feeling of patriotism. According to him honored soldiers and not wealth should be encouraged for fighting.[72] In other words, it is better to have an army who has been honoured instead of giving wealth to them.

Army vs King

The army is the force of the king and with its assistance the king is able to perform his functions. For the king, the army should work accordingly to his interest. Sometimes, the army goes against when the king is not alert. In such a situation the king loses its existence. Thus, an intelligent king never allow this to happen. Somadeva has also thrown light on those causes of army which can go against the king. He says that non-inspection of his army, taking some part of money from his salary, not paying the salary in time, not helping him in difficult situation and on particular occasions not honouring him with wealth, then the army will act against the king.[73] With all efforts, the king should satisfy the army. A lazy king who does not work and the work is done by others then he will be without wealth and army.[74]

March

When the king marches against the enemy it is the duty of the commander to equip the half army with weapons. After that the king should march forward to attack the enemy. When he is making an effort to march against the enemy then he should be surrounded by army from all sides and behind him in his camp army should be deployed.[75] The reason for this is that how a powerful king could be, his position would be like yana making him restless and then brave soldiers would attack him. When the king is far away and the enemy is moving towards him in a such situation spies residing in forest by creating smoke and fire and stirring dust and inform the march of the enemy to the king so that he can be on alert and prepares himself for the battle.[76] While entering the enemy's kingdom, a king should encamp his army in such a place which is equal to man's height, where there is a provision for small quantity of force to enter, move and leave, in front there should be sufficient space for large sabha mandala, in the middle of it he should stay and makes provisions for the stationing of his army. In ordinary places encamping the army and staying there a king cannot protect his life.[77] Somadeva emphasises that a king should avoid walking, palki or mounted in horses when he enters the enemy land because by doing so the enemy's sudden attack cannot protect him. On the other hand, he says a king should be mounted on elephant or on special vehicles and then enters the enemy's territory then he is not afraid of enemy's uprising.[78]

When the city should be enveloped? Somadeva says when the enemy is intoxicated with liquor or show laxity then the king should send the best army to the city and surround it.

Battle Formation

Management and preparation of army to fight a battle in the battlefield is called battle formation. From war perspective, battle formation is a great skill. Sometimes of this skilled a small army overwhelm a large army. A good example was of battle formation of Pandavas. Pandavas used to make new formation every day and that was the reason for their victory over Kauravas. Battle formation is of two kinds. First, when the army enters the battlefield and second when main army is kept away from the enemy and a small force is placed before the enemy.

Sukracharya and Kautilya have discussed in detail about battle formation. This has already been discussed. According to Somadeva good organised battle formation remains stable till that time when the enemy army is not visible. It means that seeing the enemy force king's brave soldier leave their formation enter the enemy's army camp and fight a ferocious battle. Such prepared type of formation becomes instable. Somadeva advises the king's brave soldiers should fight according to the rules of art of war.

Rules of War

In ancient time there were several rules of war. Battles were fought according to the rules of war. Like Sukracharya, Somadeva does not give a detailed account of these rules but, still then, he mention some rules regarding war. Perhaps, Somadeva believed in traditional rules of war. This is the reason he has not discussed in Nitivakyamitra. He writes that in a battlefield fearful and weaponless enemy bowing down at the feet of a king is a great sin. During battles captured soldiers should be freed by and should be given clothes, etc.[79]

Inter-State Relations and Diplomacy

There are few instances in ancient Indian where king ruled for a very long time. Though Ashoka, Kanishka and Samudrgupta were great emperors but they could not translate their rule into permanent empire. The reason was perhaps, lack of proper transportation. Thus, the king could not have complete control over far distant kingdoms because of these problems.

In ancient India a unique feature found in the art of war that a powerful and ambitious king wanted to become a chakravarti king. Kautilya has described a chakravati king as one whose dominion extended from Himalayas in the north to ocean in the south. Every king wanted to become a charkravarti king and hence battles were fought between them.

Somadeva has analysed three types of victorious kings (i) dharmavijai; (ii) lobhavijai and (iii) asuravjai. Dharmavijai is one who overwhelms the enemy but does not destroy his existence. Lobhavijai king is one who is eager to acquire land and wealth. After attaining them, he does not make him his subordinate but gives him complete freedom over internal matters. Asuravijai king is one who is not happy only with land and property but also kills the defeated king and takes away their wives and children. Of the three types, he places importance to the first one.[80]

Thus, the ancient India writers have discussed in detail about the inter-state relations. In Nitivakyamitra too, Somadeva has given a serious thought to it. The inter-state can be broadly directed into two parts: (i) interstate relations during peace time; and (ii) inter state relations during wartime.

Inter-State Relations during Peacetime

Diplomacy plays an important role in the conduct of foreign policy. There is a difference in the meaning and functions of diplomacy in modern and ancient times. Somadeva in his works 'Yashstilaka' and 'Nitivakyamitra' discusses diplomacy and diplomatic style with same subtlety as Kautilya. In fact, Kamandaka, Sukracharya and Somadeva Suri seem to be deeply inspired by Kautilya. The system of permanent appointment of political representatives is modern concept. There was no provision of permanent appointment of ambassadors even in the medieval period. Similarly, in India also ambassadors were not appointed permanently. Ambassadors were send to other states for some specific objectives.

Somadeva Suri observes that ambassador is the messenger of the king, therefore, he should neither hesitate in conveying the message of his master verbatim nor delay in taking up his assignments in ancient states. If an enemy sends the envoy back then the ambassador should analyse the reasons behind it. Somadeva further says, "If the ambassador get the impression that enemy is willing to wage a war with his master he should immediately apprise his king of the intentions of the enemy and send secret message to his master".[81] He prescribes the same qualifications for the appointments of an ambassador as pointed out by Kautilya. He has also compared envoys with eyes of the state because through him only king gathers information about other states.

Kinds of Ambassadors

According to Somadeva there are three types of ambassadors (i) Nishrisharth, (ii) Perimitarth, and (iii) Sasanhar. The functions of the ambassador are: to win over those persons in other kingdoms who can be divided; to create discontent towards the own rulers among those people in the other kingdoms; who cannot divided; to create divisions among the princes and other claimants of the throne in other kingdoms; to establish contact with the government of the frontier region, by entering his king's circle of states (rajmandala); to discover the spies of the enemy and to create resentment among the elements of the other kingdoms (minister, treasure, fort etc.) through desired persons.

Like modern age, in ancient time the murder of ambassadors were prohibited. Somadeva writes inspite of indulging in great crimes ambassador should not be murdered. If the ambassador takes the form of chandala even then he should not be murdered. Ambassador speaks truth, untruth, and makes pleasing and unpleasing promises. Thus, king should listen his harsh promises. When ambassador criticises his own master then he should not keep silent but should retaliate accordingly. The ambassador is the mouthpiece of king and, thus, even in grave crisis he should not be murdered.

Inter-State Relations during Wartime

Almost all ancient thinkers agreed on the principle that a king should indulge in war when all other measures fail. By other measures we mean sama, dama and bheda. By the application of these measures if a king does not get any fruitful result then he should use danda as a last resort in obtaining the desired result. Almost all the thinkers asked the ambitious king to stay away from war and use peaceful methods to resolve the issues. Somadeva too, strongly advocates this view. He states that if the king fails to solve the problem by the application of sama and dama then only he takes up arms against the enemy. According to him, it becomes the duty of minister to motivate the king to indulge in alliances with the enemy. If this fails, then the king should opt for battle. According to him, that minister and friend are like contemptible enemy because when attacked by the enemy they do not prescribed the welfare of the king by advising the king to go for alliance. Instead, ask the king either to wage the war or leave the place and move to some other place and by doing so they put the king in an awkward position.

The Theory of Mandala

Ancient thinkers believed that it was not possible to stop the war forever. Inorder to minimize the chances of war the theory of mandala was propagated by them so that some system of balance of power could be pursued. Like Manu, Kautilya, Somadeva Suri too pondered over 'mandala' as a model to maintain balance of power. In this mandala there are 12 kings. The first most important is Vijigisu who is surrounded by 11 other kings. All the kings have to conduct their foreign policy according to the theory of mandala.

The concept of mandala is, infact, the outcome of realistic thought of ancient Indian thinkers. The model of mandala was designed to ascertain the security of the state from adversary's attack as well as to neutralize their power. It also reflects the categories of power that existed in those days. Here mandala has close resemblance to Morton Kaplan model of system theory. The ancient thinkers have explained the role of the constituent states of mandala in the same manner as units of the international system have been explained by Morton Kaplan.

There are some differences in the calculation of kings as mentioned by Kautilya and Somadeva but the basic theory is the same. In the development of the theory, Somadeva has enumerated the following elements.

Udasina: The term Udasina (neutral) is related to a king who remains passive in regard to both the contending parties.

Madhyam: The madhyama king is one who exerts his influence to bring about reconciliation between two powerful kings.

Vijigisu: Vijigisu is one who is fortunate, supported by kosa, minister etc and skilled in politics. It means that a king use sama, dama and four policies and is capable of defeating his enemy.

Shatru: Enemy is also called ari. He is one who indulges in wicked act by committing crimes. Somadeva mentions the qualities of enemy king. Far distant king and close related king are friends. But this is not always possible because far distant king can be friend or foe to him. He has also discussed about three types of enemy's kings. These are natural, artificial and frontier enemies.

Mitra: According to Somadeva, a friend is one who shows kindness during peace and wartimes. Friends have been classified as daily, natural and artificial friends.

Parshnigarha: Parshnigraha is one which along with the enemy march forward for battle and after sometimes get angry and destroy the entire kingdom.

Akranda: Akranda is opposite of Parshnigarha. In other words one who helps the king to the fullest extent is called Akranda.

Asara: He is one who is enemy of Parshnigarha and a friend Akranda.

Anterdi Anterdi is one whose position is placed between enemy's king and Vijigisu and those who lives in hill.

It is clear from the above analysis that Somadeva mentions only nine kingdoms in mandala and in Kautilya's Arthasastra twelve kingdoms are mentioned.

Theory of Three Powers

A king, inorder to control his subjects, security of the kingdom and for it expansion, necessarily should have three powers. These are utsahshakti, prabhusakti and mantrasakti. Kautilya and Kamandaka have also mentioned three kinds of powers. In Nitvakyamitra, it is seen that a king with a combination of these powers is able to overwhelm his enemy. He has also discussed these powers. Prabhushakti is that where Vijigisu has a large kosa and chaturanga army. The valour and strategy of Vijigisu is called utsahshakti. Mantrashakti is strength of knowledge.[84] Like Kautilya, Somadeva has given more importance to prabhushakti. A king possessing all these qualities can defeat the enemy and vice versa leads to defeat.

Six Forms of Diplomacy

Six-fold policy is important in the conduct interstate relations. Within the mandala a king has to demonstrate his power and capabilities. By its use a king enters into mutual relations with other states. According to Somadeva, these are: (1) sandhi, (2) vigraha, (3) yana, (4) asana, (5) sanshrya, and (6) dvaidhibhava.

1. **Sandhi:** A weak king when gives wealth to a powerful king is called a sandhi. Somadeva mentions of those circumstances when a king can take the help of sandhi. Instead of giving economic punishment to the enemy Vijigisu should make sandhi with him. If the enemy in future for his own welfare does not destroy Vijigisu and Vijigisu does not destroy enemy then instead of battle he should make friendship with him. When a frontier king is powerful and he wants to acquire the land of Vijigisu then grain

produced from the land should be given to him and forge an alliance with him. This is done because once land is acquired it is difficult to get back.[85] A weak king giving wealth makes an alliance with the powerful frontier kingdom otherwise his entire wealth and kingdom are destroyed. Any two equal powers will never indulge in battles and if so both will be destroyed.[86]

2. Vigraha: To wage a battle is called vigraha. Engage the enemy when Vijigisu is powerful. Somadeva states also those conditions when a state of fighting can be harmful to Vijigisu. When a Vijigisu is more powerful than an enemy in terms of wealth and army and he is not afraid of him then he should go for war. If a Vijigisu is economically and militarily more powerful than the enemy he should fight against the enemy. It means that a Vijigisu should be aware of the situation that by indulging in battle he should not suffer losses in any form. Somadeva states once an enemy is completely subordinated he should not attack him otherwise the defeated king can take revenge by demonstrating his powers. A king should also be aware of the sweet promises of the enemy because by his diplomatic behaviour he will try to free himself and finding an opportunity he will destroy him.[87]

3. Yana: Yana is that state of affair when the enemy is attacked by the Vijigisu. Or thinking of enemy more powerful then one's own power is called Yana. When an ambitious king thinks there is a possibility of destroying the enemy's activities he should be attacked and if he had made provisions for this defence of his kingdom he should then, take recourse to yana. Somadeva makes its clear that Vijigisu should enter into war only when he is fully secured. When there is any slackness of defence or lack of proper organization in that case he should never march against him otherwise he will be met with failure.

4. Asana: Somadeva emphasizes that inorder to attack a powerful enemy when the king shows indifferent attitude towards him is called asana. In other words, a king should leave the place and go to some other places.

5. Sanshrya: Attacked by a powerful enemy and taking the shelter of some other powerful king is called Sanshrya. According to Somadeva a weak king should take the shelter of powerful Vijigisu. He should not take the shelter of a weak king because by taking the shelter of a weak king his loss will be like an uncontrolled elephant who could harm the person who try to sit on it.

6. Dvaidhibhava: According to Somadeva, fighting with weak enemy and making an alliance with powerful king is called dvaidibhava. When a king knows that assailed enemy is ready to fight with him then he should take recourse to dvaidbhibhava. He has also mentioned intelligent based dvaidhibhava. It means when a king makes an alliance with a more powerful enemy and after sometimes attack the enemy, it is called intelligent based dvaidhibhava because this will bring victory to Vijigisu.

Thus, these are important principles of six fold policy of diplomacy. Proper application of these principles strengthen the position of the king.

Instruments of Diplomacy

First of all, a well-equipped kings' army should not take recourse to war. There are other methods by which peace can be established. These are sama, dama, bheda and danda. The first three methods are to be used for before danda. In Nitvakyamitra, these four methods are as follows:

Sama: This the first method. If in any state there is no grave enmity and for simple reasons some misunderstanding takes place he should go for a conciliatory method and not directly attack the enemy. According to Somadeva there are five kinds of policy of Sama.

1. **Guna Sarkirtan**: Praising the qualities of adverse person by oneself.
2. **Sambandopakhyan**: This is a method by which makes a strong king friendship against the adverse person by applying his version.
3. **Parpokar Darshan**: Thinking good of an adverse person.
4. **Ayati Pradarshan**: One should make life successful in future by resorting to friendship by some form of demonstration. This kind of act should be shown towards the adverse person.
5. **Atmopsandhan**: My wealth can be used for your work. This is done according to one's own wishes.

Dama: For the protection of his huge property a Vijigisu should give some wealth and make the other party happy, this is called dama. Sukracharya echoes the same view. This policy can be utilized only when the first policy does not work and it is possible by the engagement of both in battle, there is loss to both the parties. In such a situation, it is useful to give certain amount of wealth to the enemy.

Bheda: This is the third method by which Vijigisu tries to sow dissension in the enemy camp through his soldiers and spies.

Danda: Under this policy, it includes killing the enemy, torturing him, seizing his wealth etc. Vijigisu should used these measures according to circumstances and situations. By the proper use of these methods victory is certain.

Espionage System

From time to time inspite of exchange of ambassadors between the neighbouring states, spies too worked and use to get the necessary information from the enemy and then deliver to his master. For any king, the appointment of spies and their use are important for the functioning of the state. Therefore, Vijigisu should sent spies to know the condition of his own kingdom and other kingdoms. Spy is an eye of the king. Their use is essential for the security and welfare of the king and the state. The qualities of spies are: satisfaction, zeal, good health, truthfulness thoughtfulness deceitfulness, graft, merchant, ascetic, mountainous tribes, attended exertion, purified witty, slanderous, sharp, cruel, dumb, deaf, blind, etc.

It is quite evident from the above analysis of diplomacy and inter-state relationship that realism characterized the theory and practice of ancient Indian diplomatic styles. Arthshastra is the most exhaustive treatise and manual of statecraft and diplomacy and his vision regarding diplomacy and statecraft appear quite realistic and they are relevant in present times as well. The ancient Indian scholars provide grading of ranks and styles of envoys. The difference in status varies with the degree of responsibility entrusted to the diplomatic agents. It is true even in modern times state appoint different categories of personnel for discharging diplomatic functions. Like Kautilya, Manu, Somadeva too had advanced concepts like befriending the enemy of a hostile neighbour, neutrality, mediation, etc. These methods are universally accepted as instruments of diplomacy even in the 21st century. In fact, a deep and extensive investigation of the Indian practice of diplomacy shows that diplomacy and interstate relationship were based on realistic principles and their impact can be seen on the conceptualization and practice of the applied framework of standard international relations discourses in the era of globalization. In modern times, an ambassador pursues the interest of his own country. His task is to implement the policies of the state vis-à-vis the state of his deputation. He adopts different types of measures such as strengthening of public relations and his participation in high level dinner parties and other public functions. He is supposed to maintain very good relationship with media. Briefly speaking a diplomat performs four basic function such as: (i) representation, (ii) negotiation, (iii) information, and (iv) observation and reporting and ultimately to defend the interest of the nation. All these functions were performed by the envoys in ancient India also. They had to follow rather than to evolve a more and extensive guidelines. Similarly rule of diplomatic immunities and privileges practiced today very much existed during ancient period.

As far as the ultimate goal of diplomacy is concerned, it may be argued that quest for power has been the most important drive among the nation states and it is the most dominant goal in world politics. During the entire period of modern history of Europe and even in the post second world war international politics, struggle for power has remained the cardinal concern of nation states. What Kautilya, Manu, Kamandak and Somadeva propounded thousands of years ago seems valid even today.

It is, no doubt, that the nature of international relations has undergone significant change in the context of changing power matrix, nuclear weapons and communication revolution. The world today is faced with the challenges of international peace and security, disarmament, terrorism and global warming. All these challenges call for lasting solutions in the interests of humanity. So diplomats in the era of globalization had to work hard to find solutions to these problems so that they can to play a crucial role for the realization of the goals of universal peace. The role played by Angad, Lord Krishna and their wisdom, vision and styles of diplomacy, as well as the practice of diplomacy suggested by Manu, Kaultilya, Sukracharya, Kamadaka and Somadeva Suri, etc. certainly provide a meaningful input for the refinement of modern trends of diplomacy.

Conclusion

The importance of Somadeva's writing in lies that he wrote in an era when the Hindu period was on decline. Harshavardhan, perhaps was probably the last king who succeeded in bringing the entire principalities of his time into a well knit single body. After that there was a dark period in the history of ancient Indian writings. In such a political chaotic situation we saw the emergence of Somadeva as an influential political and military writer. His deep understanding of Dharmasastra and Arthasastra enabled him to complete his famous work *Nitvakyamitra*. The focus of his work was on statecraft and how to protect the kingdom from external and internal threats. Here, his approach was more realistic one as compared to other ancient writers. In his own way Somadeva, too, had stated like Kautilya that knowledge was the prime requisite in world affairs, thereby placing much importance to rationality. From time to time, the emphasis on king and kingship showed that the king worked for the welfare of the subjects and his king was never a despotic king. Somadeva laid utmost importance on ministers, commander, ambassador and spies. They were there to assist the king and made all efforts to protect the kingdom from all evils so that the subject could lead a peaceful life. It was also the duty of the king to see his subject secured and this we see from his various political, economic and military policies. The thrust of his writing was on these issues because India at that time was passing through a critical phase.

Somadeva's view reflect the idea of welfare state. This could be seen from his ideas on dharma, artha, kosa and danda. State also depended upon superior arms and well equipped army. Here, he like his predecessors, has not ignored the righteous rules regarding art of war. Hence, he places more importance to dharmayuddha than other forms of warfare.

There is no denying that Somadeva's work had the impact of Kautilyan writing. There is one area where one could well co-relate Somadeva's work with Arthasastra and that is in the sphere of inter-state relations. These relate, firstly to the policy of war because of its evident risks and by contrast the advantages of policy of intrigue and politic surrender; secondly, to the justification of danda as the resort of war and of force as a preliminary to the conclusion of peace and lastly the comparative strength of the two kings in the light of war and peace. Explaining the rules of warfare, Somadeva encourages the king to fight nitiyuddha when the enemy takes recourse to evil methods. In such a situation the enemy should be dealt as the situation demands. Though political and military conditions of his time influenced him in propounding the theory of war and peace yet the impact of Kautilya was clearly visible in his writing. Still then, it is possible to assign to him an important place in the history of ancient Indian political and military thought.

REFERENCES

1. U.N.Ghosal. op.cit, p. 243.
2. B. Prasad, *The Theory of Government of Ancient India*, p. 234.
3. Ibid., pp 229-30.
4. M.L. Sharma, Rajniti in Nitivakyamitra, Bhartiya Gyanpit Prakashan, 1971, p. 12.
5. Nityakyamitra, 1,3.
6. Ibid., 2, 1.
7. Ibid., 3, 6.
8. Ibid., 5, 4.
9. Ibid., 5, 5.
10. Ibid., 29, 38.
11. Ibid., 5, 2.
12. Ibid., 6, 39.
13. Ibid., 8, 2.
14. R.G. Gettell, Political Science, p. 20.
15. Ibid.
16. See D.R. Bhandakar, Some Aspects of Ancient Indian Polity, Banaras, 1929, pp. 66-68.
17. See K.P. Jayaswal, Hindu Polity, Part II, Banglore, 1955, p. 9.
18. See. B.K. Sarkar, Positive Background of Hindu Sociology, Allahabad, 1921, Part II, pp. 34-39.
19. Kalidas, Raghuvamsham, 4.12.
20. Manu. 7.3; Sukraniti, 1.71.
21. Nityakyamitra, 17,5.
22. Aiteyra Brahmana, 1, p.14.
23. Mahabharata Shanti, 59, 144.
24. Nityakyamitra, 29, 16.
25. Ibid., 26, 16.
26. Ibid., 5, 38.
27. Ibid., 5, 29.
28. P. Sharma, Ancient Indian Political Thought and Institutions, Meenakshi Prakashan, Meerut, Delhi, 1983, p. 278.
29. Nityakyamitra, 18,15.
30. P. Sharan, op.cit, p. 278.
31. Ibid.
32. Nityakyamitra, 19,8.
33. Ibid., 27, 8.
34. Ibid., 19,11.
35. Ibid., 14, 22.
36. Kautilya, Arthasastra, 8.1.
37. Nityakyamitra, 20,3.
38. Ibid., 20, 7.
39. Ibid., 21, 5.
40. Ibid., 21, 1.

41. Ibid., 21, 4.
42. Ibid., 21, 11.
43. Ibid., 16, 26.
44. Kautilya, op.cit, BK. VIII ch. I.
45. Manu, 7, pp. 87-89.
46. Nityakyamitra, 30, 12.
47. Ibid, 30, 11.
48. Ibid, 30, 12.
49. Ibid, 30, 15.
50. Ibid, 30, 61.
51. Ibid, 30, 50.
52. Ibid, 30, 50 and 51.
53. Ibid, 30, 90.
54. Ibid, 30, 91.
55. Ibid, 5, 27.
56. Ibid, 6, 41.
57. Ibid, 9, 4.
58. Ibid, 22, 1.
59. Ibid, 22, 2.
60. Ibid, 28, 3.
61. Ibid, 22, 6.
62. R. Shamashastri, Arthasastra, translated 8th, ed, Mysore Print House, 1967, BK. X. p. 19.
63. Somadeva Suri, Nitivakyamitra, Bombay, Gantharatna Mala, 1890, pp. 83-84.
64. V.V. Mirashi, Inscriptions of Kalachuri Chedi Era, Corpus Inscription Indicarum C II, Part II, no. 45, V. 59, Ootakmund, 1955, pp. 3 and 604.
65. Ibid, Part I, no. 56, pp. 8 and 297.
66. Nitivakyamitra, 22, 11.
67. Ibid., 22, 12.
68. Ibid., 12, 1.
69. Ibid., 12, 2.
70. Ibid., 22, 14.
71. Ibid., 22, 15.
72. Ibid., 22, 16.
73. Ibid., 22, 17.
74. Ibid., 22, 18.
75. Ibid., 30, 96.
76. Ibid.
77. Ibid., 30 98-99.
78. Ibid., 30, 101.
79. Ibid., 30, 76.
80. Ibid., 30, 70, 71 and 72.
81. Nitvakyamitra, Ch. 13, p. 72.
82. Ibid., p. 73.
83. Ibid., 13, 17, 20-21, 11, 18, 19.
84. Ibid., 29, 38, 40, 36.
85. Ibid., 29, 51, 33, 65, 66.
86. Ibid., 30, 68-69.
87. Ibid., 29, 52, 54, 30, 66; 10, 141.

9

SUKRACHARYA

Introduction

Manu and Kautilya's tradition was followed by many writers on statecraft by succeeding generations. The writings of these renowned theorists extended from fifth century A.D. to the seventeenth century AD. Some of their works are well known to the student of politics and military studies, such as the Nitisara of Kamandaka, Sukranitisara, Nitivakyamritiam of Somadevasuri, Barhaspatya Arthasastra, the Agni Purana, Banabhatt's Harshcharita and so on. Treatises like Kamandaka Nitisara, Sukranitisara and Banbhatt's, Harshcharita are marked with some independent approach in the realm of military thinking. One can say that Sukraniti, like the code of Manu, does not make any conceptual contribution as compared to Kautilya's Arthasastra.

Before analyzing Sukracharya's thoughts relating to military thinking it is essential to analyse the term "niti" relating to politics, society and the protection and security of the state and kingship. Niti is derived from the Sanskrit root in which means to lead, carry, bring, convey, take, conduct or to guide, direct or govern. Niti, therefore, means guidance, direction, right course, policy, and gradually came to connote science of politics or ethics popularly known as Nitisastra.[1] According to Sukracharya, the scope of nitisastra is as wide as human life itself. Sukraniti is thus what Aristotle called the "architectonic" or dominant science. In fact, Sukracharya calls his science the science of all sciences. To him, it meant policies about friends, enemies and neutrals and also suggests international measures. Nitisastra also focuses on applied psychology because it deals with human beings interests and motives and how to cope up with the problems. It also makes observations on diplomacy, political activities and art of war so that state can be secured from various threats. He considers his Nitisastra as root of the four-fold path of life (Dharma, Artha, Kama and Moksha) and advises the king to pursue it inorder to make himself victorious over enemies, show affectionate and conciliatory towards his subjects and well should be versed in the art of statecraft.[2] Thus, the importance of the book lies in Sukracharya's comment that a person can do without grammar, logic and vedanta but not without Niti. In other words, Sukracharya discusses Nitisastra in general which could be applicable to all people and politics does not become an independent branch of knowledge for instruction of kings in statecraft but it is merged as a science of general morals.[3] Though influenced by Kautilya, Sukraniti is a very useful and highly original source of information regarding Hindu ideas on politics, sociology and war. It contains a section on the army which describes in brief the principles of war, organization, battle order, maxims to be followed and laws of war and peace.

Generally, it is assumed the work belongs to Sukracharya who is known by various names like Usana, Bhargava or Kavya. Mahabharata refers to the policy of Usana and call him by names of amilaprajana, mahayasha and yogacharya for makingan attempt to explain in terms of the essence of nitisastra spoken by Bhrama.[4] Kautilya too mentions Usana's school of polity at least seven times in his Arthasastra.[5] Yajnavalkya, too, regards Sukracharya as an authority on civil polity. Dandin in Dasakumaracharitan regards Sukracharya as one of the foremost political thinkers. Aswaghosha calls him as the builder of rajnitisastra and even Kalidas' advice is to follow the principles of civil polity laid by Sukra.

On the question of the date of compilation of Sukranitisastra there is much controversy among the scholars. According to Gustav Oppert, the work belonged to the same period which produces Manusmriti and the early epic literature. Dr. V.S. Agarwala believes that the treatise was written during the period of imperial Guptas because he finds some similarities between administrative machinery of the Guptas and the Sukranitisastra. To, Jayaswal the work belonged to 8 century AD. The problem arises when

Sukracharya conceives the idea of gunpowder which was meant to be used in war. Since guns began to be used in the early modern period hence the work has been ascribed to the 16th century. Hence V.P. Varma ascribes to late medieval period although the concepts and ideas of it is thus, clear that the book belong to earlier political institutions.[6] It is, thus, clear the writer of Sukraniti belongs to period extending from twelfth or even eleventh century to the fourteenth century. During this period the Muslim made their presence in a big way in establishing its supremacy in India.

As said earlier, Sukraniti is a work related to all the aspects of human life. Though politics forms the core of the content but Sukra has discussed about socio-economic measures which are vital to make society or state prosperous and secured them from various threats. Talking of various threats, Sukracharya gives a vivid account about the state stability and security which is depended upon the king and his council of ministers. Sukracharya has described the evil effects to the king of not following Niti viz., that the state is jeopardized both externally and internally. Hence, peace and prosperity of the state can flourish, as a matter of course, only if the relations between rulers and subjects, subjects and subjects and foreign affairs are well administered according to the precept of Nitisastra. As such, one finds that if the king and the state were to be protected then the role of army becomes important. One important aspect of his work was the importance of the development of different types of weapons. Probably, this was due to the advance of ages which had great impact on the art of war. It would be, then, in the interest of readers to know Sukras mind about the concept of state, king, army and interstate relations and diplomacy.

State

The state was always regarded as a recommendable institution in ancient India. State is composed of human being and they find protections only in and through the state. In Mahabharata, it is mentioned that without a sovereign authority to establish law and order and maintain peace, the stronger will swallow the weaker like fishes in water.[7] Kautilya also supports this view that some sort of matsya-nayaya prevails in non-states. Sukra emphasizes on rajdanda, i.e. authority of the state, because without rajdanda people can suffer and, thus, uncertainty and disorder would prevail in society. Thus, he believes without strict adherence there can be no happiness and state will suffer. It is duty of the king to see the subjects perform the functions properly and if the subjects fail the king has the power to take harsh measures. It is only through fear of punishment that the subjects follow their dharma or duty.[8] Hence, the significance of the state is related to the strict adherence to the political aspect of dharma called rajdharma in Mahabharata, dandniti by Usanas and Prajapati, Arthasastra by Kautilya and niti by Kamandak and Sukracharya.[9] As an all-pervading political instrument, state's importance lies in promoting the socio-economic and psychological needs of life. Sukra in his examination of state agrees with Bhimsa and Kautilya.

Origin of State

There was generally a tendency among ancient thinkers to accept the divine origin of the state. In Mahabharata, the state is regarded as a divine institution. The king's right to govern was partly due to his divine creation and partly to the agreement of the subjects to be governed by him in order to terminate the anarchy.[10] Kautilya also says that people themselves appointed Manu as the king and agreed to pay necessary taxes to him. Sukraniti is a practical treatise on statecraft and does not indulge into any kind of discussion on the question of the origin of state. He accepts whatever is said in Mahabharata and Manusmriti. According to Sukra, the king is created out of the eternal particles of eight gods. This has been extensively dealt in Manu's chapter. The difference between Sukra and Manu lies that the creation is not made by godhead as in Manu but by the force and energy of the austerity and penances of the king himself.[11] Sukra explicitly makes the comparison between the functions of the king to those of the gods. Mahabharata and Manu identify the king and the god but Sukra only states that king is like gods. Like, Indra, the king is an expert in collection of his own dues. The sun dispels darkness and the king is the promulgator of dharma and the destroyer of non-righteousness. Yama punishes people after death. The king is pure like fire. Like Varuna, the king nourishes the people with wealth. As the moon pleases people with its sweet rays so also the king pleases people with his good qualities and actions and he is an expert as Kubra in the preservation of a full treasury.[12]

True to his earlier view that the king is made out of the particles of gods, Sukra maintains that only that king who confirms to his dharma is to be regarded as so created. The king was not a devta but only a nara-devta. Sukraniti says, "An unrighteous king is a demon (Rakshomsah)." The Sukraniti divides kings into three classes, namely Sattvika, Rajasika and Tamasika. The king who observes his duties, protect his subjects performs all sacrifices, leads his army against his enemies, is charitable, forgiving and brave and has no attachment to things of this world is a Sattvika and he attains salvation. The king, who is reverse of Sattvika king, who is devoid of pity, is full of pride and envy and is untruthful and is a Tamasika king and he goes to hell. A Rajasa king is one who is vain, greedy, attached to the objects of enjoyment, whose deeds are different from words and thoughts, who is quarrelsome, who is fond of low company and intrigue, who is self-willed and regardless of the rules of ethics and politics, such a bad king after death reaches the state of the lower animals and of inanimate things."[13] On the basis of this analysis, one should not draw conclusion that 'Sukra does not believe in a theory of divine right about the nature of the authority of the ruler."[14]

Elements of State

The state has been defined as "a territorial society divided into government and subjects claiming within its allotted physical area, a supremacy over all other institutions.[15] In this definition, there are four elements, i.e. a definite physical territory, subjects or populations, a government and a sovereignty. In ancient India, as mentioned by Kautilya, there are seven limbs of the state. These are (i) swami (king); (ii) amatya (ministers); (iii) janpada or rastra (territory including population); (iv) durga (fort); (v) kosa (treasury); (vi) danda (army); and (vii) mitra (ally). Sukraniti also mentions these elements of the state as follows; (i) sovereign; (ii) minister; (iii) friend; (iv) treasury; (v) the country; (vi) fort and army.[16]

The Saptang theory, in modern times, has been compared to state-constitution. Sukraniti's political sovereignty and ministers are equated with government which exercises sovereign powers and imparts central unity. Forts and army are essential to safeguard the internal and external sovereignty. Treasure is important for maintaining and equipping an army, other administrative staffs and promotion of welfare of the subjects. Allies, though a new idea, could be visualised in terms of balance of modern power theory. An alliance is essential protecting and securing the above mentioned elements of the state.

Sukraniti describes the state as living organism and draws a parallel of these seven elements of the state. The sovereign or the king is the head, the minister is the eye, the friend is the ear, the treasure is the mouth, the army is the mind, the fort is the arms and the country is the legs.[17] There is another analogy drawn by Sukra related to the organic integrity of the state, i.e., the analogy of the tree and the king. The king is the root of the state, the councilors are the trunks, the commanders are the branches, the troops are the leaves and flowers, the subjects are the fruits and the land are the seeds.[18] Thus, in the organic state theory the king assumes an important place because his primary task is the integral unity of the state and also on his shoulder lies the welfare of the people. Thus, state whether in ancient India or modern India, could not survive without the king (government). This leads to further analysis of the main elements of the state i.e., king.

Kingship

The king is ranked first among the different limbs of the states. Like, Manu and Kautilya, Sukra has practically equated the king with the entire state. As Nitisastra is considered to be the spring of virtue, wealth, enjoyment and salvation, the king should carefully peruse it.[19] Since the king is the most important in the political structure of the state, it becomes essential to protect his subjects. How can the subjects be protected? Sukra answers that the primary functions of the king are to protect his subjects and punish the offenders. These two cannot be achieved without Nitisastra.

Relationship between the King and Subjects

What kind of relation should the king have with his subjects? There are some verses of Kamandaka which have been accepted in Sukraniti. These verses show the necessity of the king to fulfil the needs of the people.[20] The king is responsible for the well-being of his subjects and has been compared to moon and helmsman. He is a perfect guide without whom the subjects will get into trouble as a boat without helmsman sinks in a sea.[21] The king is like a father and his duty is to see that the subjects are endowed with good qualities. A king should be a good advisor to his subjects and teach them good lessons. It has been stressed that both the king and the subjects cannot survive without each other. Sukra also advises the subjects to show respect to the king as if "he were a second self of Vishnu."[22] They should not do anything which is against the king's interest. Similarly, Sukra advises the king to act as the lord as well as the servant of the people.[23] Hence, the duty of the king is the protection of his subjects and failing in his duty he would be a sinner. In such a case, he would not only lose his sovereignty but would be ruined and killed by gods.

The relationship between the king and his subject is also determined by fear of punishment when the subject fails to perform the function peacefully. Sukraniti says "though fear of the punishment meted out by the king, each man gets into the habit of following his own dharma or duty. Without strict adherence to one's own walk in life there can be no happiness.[24] The king should make the subjects acquire the habits of performing duties by the use of his terrible sceptre. And he himself should practice his own religion otherwise his influence will be on the wane."[25]

The importance of the king and his authority is here brought out by the fact that it is the sceptre that would keep each subject to perform his duty well and thus the people would be happy and prosperous and such the state will be protected by the evil designs of others including the enemy. Sukra believes that no good could accrue to a king through punishment of his own subjects as that will lead to destruction of his fame, wealth and virtue. He, further, mentions that satyayuga had no danda as the king then worked according to dharma. In the treaties, vices had crept in the society and punishment became important to keep the society intact. In dwapara, danda was practiced because of the decline of virtue among the subjects. In kali, the poor people suffered heavily because of the wickedness of the king. Sukra has boldly emphasized that the king is the maker of the age as the promulgator of duties and sins. The fault lies neither to the age nor to the subjects but to the king.[26]

Sukra further states that there are eight functions of the king: punishment of the wicked, charity, protection of the subjects, performance or rajasuya and other sacrifices, equitable realisation of revenues, conversion of princes into tributary chiefs, quelling of the enemies and extraction of wealth from land.[27] This shows that even the independent princes were to be made dependent and tributaries. The king should always try to be an overlord by conquering kingdom and establishing an empire over them so that their chief may be paid to pay homage. The advice is imperialistic as we see in the writing of Kautilya. Furthermore, following the tradition of Manu, Sukra also mentions four means or upayas to be applied by the king in order to govern successfully. These upayas are sama (peace), dana (purchase), danda (penalty) and bheda (partition). These methods are to be adopted by the king only when the strength of the enemy is known. But so far as one's own subjects are concerned Sukraniti denies the use of bheda and danda as the state is likely to be ruined through punishment and partition of the subjects. Only sama and dana have been prescribed in the case of own subjects. It is clear from Sukraniti that the welfare of the people is the most important factor of state's prosperity and stability.

Relationship between the King and Ministers

Sukraniti also mentions the relationship between the king and ministers in conducting the affairs of the state. Minister is the eye in the body politic of the state. The king is placed in an uncomfortable position without his ministers who advise him from time to time. It is said that even the king fears the control of good ministers. The ministers also act as a check to the arbitrary powers of the king. It appears that the influence of the ministers over the king is essentially moral and not a constitutional one. Normally, the king is supposed to respect the opinions of the ministers and in return the ministers worked for the betterment of the king and his state. The ministers are supposed to be bold and intelligent so that they can check the king from doing wrong acts. Sukra, thus, was in favour of strong and capable ministers who could put a check on the arbitrariness of the king.

So far the strength of the council of ministers is concerned the Surraniti mentioned a council of ten ministers. There are: purohita, pratinidhi, pradhana, sachiva, mantra, pradvivaka, pandita, sumantraka, amatya and duta. The responsibilities of these councillors are to work for the betterment of the administration and subjects. The king is, therefore, advised to appoint proper and able persons as ministers otherwise, they would be the cause of the ruin of the state and the king. Moreover, honest and bold ministers could be far away from vices and would try to take correct decision if he fails to do.

After detailed discussion of king and state, Sukracharya envisages the force of use in bringing stability and prosperity in state. Force here means the military methods by which a state can be secured from internal and external threats. Hence, it is important to examine Sukra's view on war and the organization of army.

Art of War

Sukracharya defines war 'as the affairs that two parties, who have inimical relations with each other, undertake by means of arms to satisfy their rival interests'. In modern international law, "war is a contention between two or more states through their armed forces for the purpose of overpowering each other and imposing such conditions of peace as the victor pleases." Mao-Tse-Tung describes war as "the form of struggle between nations, states, classes or political groups for the purpose of achieving victory for themselves." In any case, war involves the use of force for the realisation of political objects or object; to conquer or destroy the enemy's armed forces; to occupy the enemy resources and to gain public support. All these features of war are also found in the writings of Sukracharya.

Sukra mentions three kinds of war. These are daivika (where charms or mantras are used), the asura (where mechanical devices are used) and the manava (where sastra and hand are used) and this fight of one with many, of many with many or one with one, etc. The daivika, asura and manava are termed on the same basis in another passage as uttama (fought with charmed instruments or mantras), vsadhyama (with mechanical instruments), Kanishtha (with weapons) and adhama (with hands) as different kinds of war. It seems strange that Sukra supported asura war over daivika war. Generally asura or demonic war is inferior to human one but accepting asura war as better than human war, Sukra's concept of war was different from those thinkers who propagated righteous war as true war.

In ancient India, inspite of rich culture and tradition war was given importance to protect these traditions. But Sukra approach was quite different and he regarded war as the last resort. Before going to war, Sukra advises the king to adopt sama, dana and bheda in order to overpower the enemy. The danda was to be employed when all the other means failed. War as last resort was well acknowledged by Karl Marx, V.I. Lenin, and by the countries in recent times. The best example was the USA attack on Iraq in 1991.

Seasons of War: Sukra envisages that the king who wants to fight should carefully consider the season, the region, the enemy's strength, own strength, the fourfold policy and the six elements of statecraft. He considered autumn, hemanta (October and November) and winter were the best seasons for waging wars. The spring is good but the summer is the worst season. In the rainy season, war is not at all appreciated then peace is more desirable. Even Abul Fazal quotes the Hindu ideas in the following words: "if he finds it necessary to attack the enemy, he should invade his country during the time of harvest." This was also Napoleon's military method.

Even if otherwise the time is good for war, i.e. if the seasons and weather are favourable, the king should not necessarily be tempted to go for war. He should first look at the pressing problems inside the kingdom. Sukra, then advocates the theory that the man is the maker of his time, he should carefully analyse his own conveniences and never be influenced by purely physical considerations of seasons, weather, climate etc.

Army

Sukra mentions that war is fought by armies and army is considered as one of the seven elements of the state. The state and its subjects are fully secured when there is a military system and it is the duty of the king to maintain a good army. The army, to Sukra, is the group of men, animals, equipped with arms, missile etc. Without the army one cannot conceive of state, wealth and prowess. While emphasizing the importance of army, Sukra says "what mind is to the human body, the army is to state." What is important for the state? Sukra says that strength of the body, strength of valour and prowess, strength of life are important for the protection and stability of the state. One who has all these qualities is equivalent to Vishnu.[28] Without the army one can overcome even an insignificant enemy.[29] The army is the chief mean of overpowering the enemy and so the king should carefully maintain a formidable army.[30] Sumanth mentions army is subpart of treasure and not a separate part of state. On the other hands, Kautilya and Sukra regard it as one of seven elements of the state. So importance Sukra gives to army that he allocates 50% of the total revenues on the army.

The Sukraniti gives figures relating to the proper apportionment of military expenditure to be incurred. We may take the figures as roughly representative of earlier times as well. The following estimate gives, according to the treatise, the proportion of the sums which may be spent on the various items of civil and military expenditure by a State with an income of 100,000 karsas. The allocation of funds provided by Sukra has been shown by the Tables given below:

Table 9.1: Items of Expenditure (annual)

1.	Personal	18,000 karsas
2.	Clerical	1,200 karsas
3.	Council (perhaps executive)	3,600 karsas
4.	Family	3,600 karsas
5.	Men of letters	2,400 karsas
6.	Cavalry and infantry	48,000 karsas
7.	Elephants and other service	4,800 karsas
8.	Savings	18,000 karsas
	Total	99, 600 karsas

Source: Sukra IV, 7, 27-29.

Further, we are told that a State which has the same annual income is expected to have the following military establishment:

Table 9.2

1.	Reserve Soldiers	100
2.	Infantry men with guns	300
3.	Cavalrymen	80
4.	Chariots	3
5.	Guns	2
6.	Camels	10
7.	Elephants	2
8.	Bulls	16
9.	Clerical Staff	6
10.	Higher officers	3

Source: Sukra IV, 7, 24-26.

Classification of Army

Most writers have mentioned six types of army: maula, bhrtya, sreni, mitra, amitra and atvika. But Sukra classifies army into two categories. It either belongs to oneself (swiya) or to the allies (mitra); from the points of view of recruitment it is either standing army (maula) or a new recruited one (sadyaska). Again the army is trained or untrained; officered by the state or not officered by the state; equipped by the state with arms on supplying their own arms and ammunitions, bringing their own vehicles or supplied vehicles by the state.[31]

It would thus, appear that the maula army would correspond to the permanent standing armies of the kingdom and the sadyaska was new recruited army which was enlisted by the methods of conscription or voluntary services. But, the idea of military conscription was not supported by other military thinkers in ancient India. Similarly, the military thinkers mention bhtra, bhrtaka or bhrtya (all meaning paid ones) in place of sadyaska. An interpretation could be that maula constituted persons belonging traditionally to warrior class serving the state and the sadyaska were the newly recruited warriors. It would be correct to assess that Sukra's classification of army was based on recruitment, command, accoutrement, vehicles and training.

Furthermore, Sukra classifies his army according to the nature of movement. It is of two kinds (1) swagama (that moves without any help – infantry) and anyagama (pulled by vehicles) – chariots, horses and elephants. He also refers to satrubala. It is an army comprising soldiers which are left by or captured from the enemy and becomes a part of one's own people. But both of these kinds of troops are said to be of no much use and should be regarded as inimical.[32] In a similar way, from the point of view of usefulness the army is called sara or asara.[33] Thus, the classification of the army appears to be systematic and well regulated by rules and is the real strength of the state.

Division of Army

In ancient India the army comprised of infantry, chariot, elephantry and cavalry.[34] Some military authorities mentions six and sometimes even eight parts.[35] Sukra follows the ancient tradition and mentions four–fold army.

These were infantry, chariot, elephantry and cavalry. The first is called swagama while the remaining three anyagama.

Besides these four divisions, Sukra recommended camels, bulls and canons as further divisions of army. His advice to the king is that he should have preponderant majority of infantry, a medium number of horses, a small number of elephantry with equal number of bulls and camels. What should be the ratio of these divisions? Sukra states a king's infantry should be four times than cavalry, bull one sixth of his horses, camel one eights and elephants one fourth of camels, chariots one half of the elephants and cannon twice the chariots.[37]

Army Strength

The king inorder to make the army strong, efficient and well versed in the art of war should be imparted proper training and regular exercises. He also advises the king to practice military arrays and archery and parades with the troops. He also suggests the strength of the army should be increased by hand to hand fights, by regular exercises, parades and nutritional food. The valour and prowess to be increased by development of weapons. The army strength also depended upon the payment made to the soldiers. The payment could be in the form of regular pay and rewards. The army which is well rewarded by gifts and honours will remain loyal to unrighteous and low born king. All these increase the military spirit and morale among the soldiers so that the soldiers can fight battles courageously. By doing so, his empire will be safe. So long the kingdom lasts so also the king is alive. This is done by intelligence network involving even people who are well informed in Sastra.

Officers and Status

Sukra recommended hierarchy to be maintained among the officers. The king was the supreme commander of the army, sachiva, in his council of ministers, was his chief advisor on military affairs and commander of the army was perhaps senadhipa.[38] He followed the Mahabharata's tradition. The smallest unit was known as patti. But in Sukraniti its strength differs from that mentioned in Mahabharata. In Mahabharata there is reference of patti that comprised of 55 soldiers but Sukra recommends only 5 or 6 soldiers to patti. The chief of patti is called pattipal. As compared to other works, Sukra mentions different grades of officers.

There is reference of gaulmika who was the head of 30 soldiers and the chief of 100 soldiers was known as satanika. The best anusatanika, senani and lekhaka were appointed over every 100 soldiers as heads for various kind of functions. The officers-in charge of 1000 soldiers was called sahasrika and the ayutika was head of ayata or a division of 10,000 solders. An ayuta is also mentioned in Mahabharata which comprised of the same number of soldiers. Satanika was responsible for regular training of soldiers to make them perfect in performing the military functions. Senani, gaulmika, pattipa and lekhaka were responsible in giving directions to guards, inspecting night duty, conducting the rotation of watchman on duty at night and maintaining accounts and different types of reports of the soldiers respectively.

Sukraniti goes one step ahead by mentioning different grades of military officers. Satanika, the commander of 100 was equal in status with an officer known as nayaka, master of 10 villages. One could also draw inference from the above that nayaka of 20 horses or elephants much have enjoyed the same status. Samanta, the ruler of 100 villages, was equal to sahasrika, the commander of 1000 troops. It seems that both had vehicles which were driven by one horse and 10 attendants. Manadalika, the ruler of 1000 villages and ayutika, the commander of 10,000 troops, were equal in status and had vehicles which were driven by men or two horses or by an elephant with 20 attendants. Asapaka or swarat, the commander of 10,000 was equal in status with that of the commander of 50,000 troops. All these indicate that the officers occupying higher status are found in modern times also. In modern times, the head of these troops were more or less in the hand of civil officers. It is quite strange that it was only Sukra who discusses the equality of status between military and civil officers.

Military Leadership

The essential qualities of military leadership are intellect, military education, exceptional qualities of character, strength of mind and soul, deep devotion to duty and all round personality. A superior military leader requires these "to overcome the frictions inherent in war' and to reach "heroic decisions based on reason" as the truest mark of superior leadership.

Clausewitz has also emphasized that "no military leader has ever become great without audacity." Similarly, Sukra stresses agility, elasticity and flexibility on the part of a general. He must be stimulated by a powerful emotion. Motivation, thus, plays a great part in leadership. History is full of examples where commanders have showed their skill and leadership. Caesar was ambitious, Hannibal was known more for hatredness toward his enemy, Frederick, the Great stressed on pride in a glorious war. Babar was a military genius of first rank.

Unity of Command

Sukra attaches much importance on principles of the unity of command in army. Such principle would cause disaffection among enemy's soldiers and then the task to subjugate them would be easy.

The king is advised to rely more on maula troops in the times of war. He is also advised to reward his soldiers by giving awards and booty so that the soldier can prove to be more faithful to him. Sukra also mentions of policy towards army which is to be followed by the king in order to maintain loyalty and unity. The king is also expected to analyse the causes of alienation and disaffection of the army belonging to himself or the enemy. Sukra believes that the alienation among soldiers in generally caused by harsh words, cut in wages there as threats and longer period of works and life in foreign lands. Hence, the king should refrain from these things. He suggests that the soldiers must be controlled by special methods and never by fines.[39]

Weapons

It is quite strange that Sukra does not go in detail about various types of weapons except gunpowder. It is these weapons that most scholars correlated his work to, a date not earlier than 13th century.

Sukra mentions two types of weapons called astra and sastra. Astra is one which is thrown or cast down by mean of charm, machines or fire. Sastra is any other weapons, e.g. sword daggar, kunta, etc.[40] Astra is of two kinds: charmed or tubular. The king who desires victory should use tubular where the charmed does not exist. It should be, then used together with sastra.[41] People expert in military instruments should know of diverse agencies named astra and sastras varying according to long or short size and the nature and mode of the sharp edges.[42]

Thus, the classification, made by Sukra is simple as compared to other writers. Other works like Nitiprakasik, Agnipuran and Vishnudharmottara have divided these in four or five categories. These are mukta (thrown), amukta (not thrown), mukta-mukta (thrown an received both), and yantra-mukta (thrown by missiles) or panimukta (thrown by hand), and niyudha or bahuyudha (wrestling etc).[43]

Sukra further divides astra into sub categories: the mantrika and nalika. The nalika (tubular or cylindrical) astra is known to be of two kinds according to large or small size. The small nalika is the cylindrical instruments to be used by infantry and cavalry, having an oblique (horizontal) and straight (perpendicular) hole at the origin (breech). The length is of five vitastis (two cubits and a half). It has a sharp point both at the forefront (muzzle) and at the origin which can be used in marking the objective and has fire produced by the pressure of a machine. It contains stone and powder at the origin. It has an inside hole of the breadth of the middle finger, holds gunpowder in the interior and has a strong rod.[44]

The large nalika is one which has a post or wedge at the origin or breech and according to its movements can be pointed towards the aim. It has a wooden frame and is drawn on carriage. If it is well used, it will lead to victory.[45]

Sukra also speaks how gunpowder is prepared? Suvarcisalt-five palas, suplur-one pala and charcoal from the wood of arka snuhi and other trees burnt in a manner that prevents the escape of smoke (in a closed vessel) – one pala, have to be purified, powdered and mix together, then soaked in the juices of sncehi, arka and garttic, then dried by heat (of the sun). The material is finally powdered and looks like sugar crystals. The substance is gunpowder.[46]

Six or four palas of survarci salt, sulpur and charcoal remaining the same, may also be used in the preparation of gunpowder. Charcoal, sulphur, survarci salt, mixed with haritala, lead (sisa), hingula, kantisara lauha (iron filings), camphor (karuc) jatic, indigo extract (niti) juice of sarala trees, less or more are powdered to prepare various types of gunpowder. This when burnt gives flames like moon light.[47]

How shells/shots (topas) are made? The shell made of iron balls in which other smaller shells are filled. For guns, shells are made of lead or other metals. The shells for guns are made of steel or other metals. The guns have to be cleaned properly and the gunman should stay by its side. Shots are fired when the gunpowder is pressed tightly inside the original chamber with the rod. After then strong shells or metal balls are placed. Gunpowder then filled through its side hole and fire is put on. By this, shells or balls will burst towards its pointed direction.[48]

Thus, one can say that the use of wooden barrel for cannon was somewhat new in its antiquity. It is reported that the use of bamboo filled with gunpowder was used in China in 1132 AD. Metal barrels for hand guns were also used in China by the end of 13th century AD. The question is what is survarci salt? In Kautilya Arthsastra it is referred as materials for purification of gold and used as oxidizer which readily burns. There is no doubt that the saltpetre was used to prepare gunpowder. This salt is taken otherwise as potassium nitrate (KNO_3), Sodium nitrate ($NaNO_3$), Calcium nitrate Ca $(NO_3)_2$ or chlorides and sulphates of sodium and potassium. The calcium and sodium salts are hygroscopic in nature and becomes useless in humid atmosphere. The potassium nitrate is a white non-hygroscopic salt. It has three active oxygen atoms per molecule making it an ideal oxidizer when hot. A standard formula for gunpowder is KNO_3, salt (75%), sodium (10%) and calcium (15%). It appears that Akbar had knowledge of the properties of saltpetre which was used for generating heat (as explosive) as well as coolant for drinks and producing ice.[49] One can say that Akbar was aware of both hygroscopic and non-hygroscopic nature of various salts.

Sukra also mentions other sastras like arrow, club, different forms of spears or darts (pattisha, prasa and kunta), sword, disc with sharp teeth, nose, finger nails with armours and helmets.[50] Sukra also prefers the use of mantrina astras but the author does not discuss in detail.

Military Operation

After discussing the weapon system as laid down by Sukra, he says that now the army is fit to wage battle or war. In Sukraniti the term yana is used which means an aggression on enemy. Sukra mentions five kinds of operation: vigrihya, sandhaya, sambhuya, prasanga and upekshya. In vigrihya, the army moves slowly over shadowing group of enemies. To Sukra, vigrihya is an expedition in which one's friend fights with the enemy's friends on all sides and his own army moves against the main enemy. Sandhya is that kind of operation when alliance is made with some supporters of the enemy. The sambhuya operation is made by the king based on the contribution made by the resourceful and skilled feudatories. In prasanga operation the army attacks a certain enemy but incidentally moves to fight with another enemy. The last operation, upekshya, ignores the enemy and then returns back facing adverse results.

Bases

Sukra refers to various positions while fighting a war or battle. The term asana is mentioned as one of six-elements of the state. Sukra defines it as a position from which one can be protected and the enemy is destroyed.[51] His advice is to deploy the troops at such places from where enemy's army could be defeated with astras. The asana should be taken in such a way that supply line to the enemy could be cut. From its maneuver, one is advised to destroy those people who help the enemy by carrying wood, water and provisions.[52] Sukra also refers sandhaya asana or a ceasefire or cessation of hostilities. This happens only when both the enemy and the aggressor are tired of fighting with no result. They, then stop the war but they do not retreat from their respective positions. Sukra also mentions asraya. Asraya means taking shelter with friends, relatives and kinsfolk or with paid friends or sharer in the benefits of victory.

Battle Order

Formation of battle array is regarded an important feature of traditional military science in India. This is also relevant in modern time. The strategists command a mastery over battle order according to modern tactics and methods of warfare. A commander's job should not only show courage in battles but also prepare for better strategy in the formation of battle order.

The commander has been advised by Sukra to employ different forms of battle orders but also on the way whenever difficulties arises through rivers, hills, forests and forts etc. He was not in favour of single battle order. The commander should be capable of forming one, two or more of the vyuhas or a mixture of them inorder to keep intact the strength of troops as well as the topography, roads and battle fields and array of the enemy.

Sukra, in certain situations, mentions the significance of some of the vyuhas. If there is a danger ahead the commander should march in great 'makara' or crocodile array, or the shyena or bird array which has two wings or the 'shuchi' needle array which has sharp mouth.[53] If danger is from behind then the following arrays are to be followed, shakata (carriage), sarvatobhadra (octagonal), vajra (thunder) array, chakra (wheel) and vyala (snake) array. The array should also be determined according to the nature of the region in such a way as to pierce the enemy's array.[54]

Sukra further says that the wise king will always devise diverse forms of battle array for horses, elephants and footsoldiers. He should order the soldiers aloud by signs of battle order from a station on the right or left, in the centre or in the front. Once the orders have been heard the troops are to carry out the instructions. He lays emphasis on grouping, expansion, circling, contraction, strategic movement, rapid march, backward movement, separation in parts, standing in serial rows, holding the arms and weapons, fixing the aim and striking the objective. During operation, the army can move forward, backward or sideways.

Sukra has also mentioned about the placement of the different types in the battlefield. After creating disaffection among the enemy's troops one should start fighting by placing both types of fire arms in the front, the infantry at its behind and the elephant and cavalry in the wings. He suggests that the first attack is to be made by the commander's half forces in the front and the wings so long as the region is favourable for warfare is not acquired. The war should then be taken with troops successively by ministers and the king.

Sukra warns the king that deserter troops from the enemy should not be placed near the main army. So far the troops of the friends are concerned, they may be placed in the front, at the rear or the wings. Moreover, the king should not make use of untrained, inefficient and raw recruits in the actual war. They may be given other works. During operation, the king is advised to wage battle even with a small enemy by large number of forces.

Principles of Ethics in War

War is often cruel and destructive in nature. The ancient Indian thinkers have tried to make it "as human as possible" by framing a code of conduct agreed upon between the parties. In Mahabharata there are references of fighting a dharmayudha. Manu, Gautam, Yagnavalkya and some others subscribed to these rules.

Sukra, too, like his predecessors, mentions certain rules to be followed by both the warring parties. Soldiers and weapons equally matched should be used against each other. For example, the horsemen was to be attacked by the kunta sword, the charioteers and the soldiers with the elephant by arrow, the elephant by elephant, the horse by horse, the chariot by chariot, the infantry by infantry, one by one, the weapon by weapon and the astra by astra.

There were also rules regarding the soldiers fighting in the battles. Sukra says that soldier is not to be killed who is lying on the ground, who is deformed and in posture of surrender, who is seated with hair disheveled, who is asleep or naked or unarmed, who is drinking water, taking food or busy with other matters and who is terrified and retreating. In similar ways an army should not kill the old man, the infant, the women as well as the lonely king.[56] But Sukra allows the killing of infant or a Brahmin if any of them is coming upon with weapons in hand or with murderous intention.

Sukra did not appreciate those soldiers who in order to save their lives flee from the battleground. Such soldiers are deemed to be dead and are liable to suffer the sins of the whole kingdom.

Despite these rules of dharmayudha, Sukra also prefers Kutayudh against powerful enemy. Sukra gives the examples of Rama, Krishna, Indra and other gods who over-shadowed the enemies by using deceitful tactics. It is clear that Sukra, being a practical thinker, was an advocate of expediency and diplomacy. Therefore, so called irreligion immorality, inconsistency and untruth, which are the features of Kutayudha should not at all deter the king from course of action that would lead to the desired object.[57]

As stated earlier, a strong king is equal to Vishnu. The individual soldiers must possess moral qualities like courage, audacity and self-sacrifice. As Jai Singh wrote to Aurangzeb after the conquest of Purandar and on the eve of the Bijapur campaign: "there is much work here for intelligent and servicable people imbued with the zeal for work, honesty and loyalty." Mere courage is not enough but more important than it is the morale, which produces certain physical effects. As early as 5th century BC Confucius in China emphasized the value of morale in an effective tone: "with walls of great height, with moats of great depth, with arms, both of offence and defence, trenchment and mighty, with great stores of rice and other food, the city is surrendered and abandoned." This "is because material advantages do not compensate for the absence of the spiritual union of man".

Fortresses

Fort is regarded as one of the seven elements of the state. A small but complete section has been devoted to fort by Sukra. He says that fortresses are made inaccessible through ditches, thorns, rocks and deserts. Like Manu, Sukra mentions six different types of forts. According to him parikha fort is surrounded on all sides by great ditches and the parigha is protected by walls of bricks, stones and mud. Vana or forest fort is one which is encircled by huge thorns and clusters of trees. The dhanwadurga (desert) is situated in such a place where no water is found. On the other hand, jaladurga (water) is that which is surrounded by great sheets of water. The giridurga (hill) is described as that which is on high level and is supplied with plenty of water. The sainyadurga (troops) is that which is defended by heroes well up in vyuhas or military tactics and hence is impregnable. The sahayadurga (help fort) is known to be that which belongs to the valorous and friendly kinsfolk.[58]

Of all these kinds of forts, Sukra regards desert fort as superior to the parikha. The sahayadurga and sainyadurga are the ornaments of all fortresses. Without these, the other forts are of no use to the king. In other words, the fortress with soldiers is the best of all and the others are mere help to this. Therefore, the king should always keep this fort. Sukra further says that the king having forts with troops can survey the whole earth but to have every other kind of forts without troops is tantamount to imprisonment. Sukra emphasis is on mutual dependence among different kind of forts and their resourcefulness. Hence, the king is advised to have all these forts well provided with all types of materials which are necessary for war.

Diplomacy and Inter-state Relations

In ancient Indian, almost all thinkers gave importance to the principles of diplomacy and inter-state relations. Like Manu and Kautilya, Sukracharya has also discussed this aspect of statecraft in much detail.

In discussing the most important objective of foreign policy, the king has to win overcome powerful friends so that his state is secured and the enemy strength is weakened. The term 'friend' has also been referred but Sukra calls it "ear to the state". Manu has given more importance to friend then gold or land and in Kautilya's list friend is placed at number three. So far Sukra is concerned, his approach is quite similar to that of Mahabharata. He says that all kings are unfriendly and secret enemies of powerful and brave soldiers as they themselves are covetous of kingdom. In this sense, he believes in the dictum there is nothing like permanent friendship in statecraft. Therefore, it is necessary for a king to make friend or enemy.

Mandala Theory

It was Kautilya who was the originator of this theory. Earlier thinkers like Manu and Kamandaka too mentioned the theory of mandala. Even today, we find modern governments have interpreted interstate relations as visualized by Kautilya. The theory imparted a very symmetrical relationship resulting from the quest of suzerainty and the consequent need of astute diplomacy and alliances.[59]

Sukra envisages the circular inter-state relationship. He places no importance to madhyama king but in Kautilya's theory the madhyama has an important role to play. Sukra's classification is restricted to only certain categories: ari (enemy), mitra (friend) and udasina (indifferent and super). In Sukra's plans kings are placed in the following order: in all sides first enemies, then the friends and last the udasina. It appears that Sukra might have accepted geographical basis of friendship or enmity as well as the earlier practice that enemy's enemy is always a friend. To Sukra, the nearest state is the greatest enemy but these gradually, receding from king's neighbourhood are lesser enemies successively.[60] Similarly, the king's duty was to check the ministers and servants who were very close to him.

But Sukra, like Kautilya, accepts that the immediate neighbour in front should be assumed to be enemy. There are good examples from modern history to illustrate this view. Prior to the second war the relationship between Germany and France, Poland and Russia and China and Japan were inimical because of their contiguity. This hampered their interests from time to time. After the Second World War the proximity of border between India and Pakistan, India and China, Russia and China and Iran and Iraq are good examples to prove the above theory as envisaged by Kautilya and Sukra.

Sukra, like Kamandaka, defines an enemy as the man who is a rival claimant of exclusive ownership over the same matter and is a destroyer of one's interests.[61] Sukra, further analyses friends and enemies, based not on geographical situation but on their actions towards each other. Such types are of four kinds; (a) one who does not make others do; (b) approves of and helps in things that are respectively beneficial and harmful.[62]

Sukra was also well aware of the importance of the theory of balance of power. His advice to the king is to adopt the policies in such a way that enemies neutrals and even friends can never go beyond him.[63] He never accepted the idea of a friend king becoming more powerful than oneself. Even among friends he wants to maintain a balance of power. The king powerful friend has been asked to encourage the less powerful friend and to control more.[64] Thus, Sukra does not propagate the idea of a powerful friend becoming more powerful and hence a balance has to be maintained among them.

The Four-fold Policy

The ancient Hindu thinking laid much emphasis on sama, dana, danda and bheda. These fourfold polices are applicable even today in international politics. To Sukra, sama means praise or appreciation, dana is to give money, land, award, etc., bheda is to sow dissension among friends and danda means punishment. Sukra believes that this policy should be used carefully. He states that peace, purchase, partition are to be applied separately to each of the following classes of relations, i.e. the friends, relations, family, subjects and enemies according to one's own interests.[65]

Sukra mentions ways and methods to obtain the objective of the fourfold policy. Its application to the enemy is obviously different. Sometimes the relationship of mutually not injuring but helping each other in time of crisis is called sama. Dana is applied to an enemy when the powerful king is pacified by paying tributes or annual incomes from different sources of land according to their strength and status. Bheda is a method by which a king separates his enemy from others and makes him powerful. Danda is applied to the enemy who attacks him with a powerful army and there are no chances of ceasing the war.

Sukra applies the fourfold policy differently in different situations. He outrightly rejected the application of bheda and punishment to friends. Sama is to be adopted towards friends and not otherwise. A similar approach is to be adopted by the king when he is dealing with his own subjects. One should never rule his own subjects by the policies of bheda and punishment but by sama and dana. If so happen seeds of dissension will be sown in the state and the king's powers will be hampered and then he will have to face internal problems of security. As such, the state will not be secured. Thus, the general principles of fourfold policy are meant to the enemies. The practicability of Sukra's suggestion is self-evident from the fact the each policy should be used properly with regards to friends and foes. Not only this, Sukra suggests a king should apply this fourfold policy upon the subjects of the enemy. The disaffection among the enemy's subjects will boost the spirit of the king and will be a cause of success. The enemy's subject should be subjugated by offering sama and dana. This also shows Sukra military insight in dealing with righteous friends and foes. Sukra, like Manu and Bhimsa, makes it clear danda should be applied as the last resort when other methods fail. Thus, Sukra's approach is quite modern in terms of diplomacy where state tries to avoid war as far as possible.

Over all, it seems that Sukra has attached great importance to bheda. The sovereign king is asked to adopt bheda and samasrya (refuge with a stronger one) in order to overwhelm the enemy. Means could be rivalry or conflict between senapati and councilors of the enemy and strife among their subjects or women. It is said that the weaker enemies are to be won by wars, the stronger ones by honours and presents and very powerful foe should be subjugated and humiliated. Equal powers are to be won by alliance or friendship. The enemy could be subjugated by causing disaffection among the soldiers. It is, therefore, clear from this that Sukra's idea of inter-state relations was quite bold and, to a certain extent, he differed from his predecessors.

Sadgunas: Sadgunas mean alliance or treaty, war, expedition, stationing, refuge and duplicity. These attributes are important aspects of a state in the formulation and execution of foreign policy with other states. These are essentially meant for preparation or fighting a war and hence the last five have been discussed in detail in this chapter.

The important question, then, arises what is sandhi (alliance). Sandhi is not necessarily a part or cause of war. It belongs to the realm of diplomacy. Even Sukra has given enough space for discussion. He argues that a powerful aggressor never returns without getting something as there is no other form of alliance but gifts.[66] Sukra's definition of treaty is one by which a king by his act makes powerful enemy his friend. His advice to the king is not to indulge in any kind of war with powerful enemy. Sukra explains this by a simile. He says that prosperity never desserts a man who bow down to the powerful at the proper time, just as river never leaves the downward course.[67]

On the question of giving gifts to the enemy who has concluded a treaty with the king, Sukra says it should be determined by their status and power. In this, it is also included their services, wealth and even daughter as gifts. Sukra does not favour the idea of religious or racial relationship becoming more important than political necessity. It is the political power that governs the diplomatic behaviour of the king.

Sukra's approach seems to be more realistic when he gives a suggestion to the king to make treaty with the powerful one. There is a saying that in every alliance there is horse and a rider and Sukra does not object to being even a horse if the other side is more powerful. However, the king is warned never to trust enemy even if the later had entered into an alliance. Here, an example can be cited of the King Indra who killed vitra in days of yore during the peace time. On the whole, Sukra's idea was more close to Kautilya. Like Sukra, Kautilya also advises the king to accept the most humiliating terms in time of crisis from a more powerful enemy and to wait for his opportunity to free himself from this servitude.[68]

Attitude Towards the Defeated Enemy

Sukra appears to favour the idea of lobhavijaya. He states that the victorious king should realise revenue from the defeated enemy according to his capacity. Sometimes, keeping in view the merit of the case, he could make annexation of a portion or half or the whole of the conquered state to his own territory.[69] On being victorious, he should change the member of the Council of Ministers and other important officers of the defeated country. This is done essentially to make his presence felt by the subject of the defeated country. Moreover, Sukra does not believe in killing the enemy or his family but provide better provisions for living a respectful life and also grant maintenance allowance to the defeated king and his family. Sukra even offers ¼ of the conquered territory to the qualified son of the defeated king otherwise he will get only thirty second part of it. This shows Sukra adopted a soft approach towards the defeated enemy.

The king was permitted to retain the treasure of the conquered state for his own enjoyment. This means that the treasure won by the king was his own property while the taxes and other income from the conquered territory were regarded as revenue of the state. Sukra also instruct certain rules to regulate the king's behaviour towards the employees of the defeated one. He should not destroy the army of the enemy's friend even if they are in danger.[70] By doing this, he tries to sow dissension between enemy and his allies.

Sukra further says that the king has to act according to certain principles laid down by law. Here he highlights both the aspects of the king the material and the human interests. This shows that Sukra follows the Indian tradition, which does not permit the annexations of territory and killing of the defeated kings and their families but takes measures to reinstate them if their behaviour is good. It has been well pointed by Sulaiman about the peculiar human method of conquest. He has observed "When a king subdues a neighbouring state, he places over it a price belonging to the family of the fallen king who carries on the government in the name of conqueror. The inhabitant will not suffer it to be otherwise.[71] Sukra in a clear tone speaks that the king should maintain the dispossessed princes for the display of his own majesty by the bestowal of honours if well behaved but punish them if unliked.[72]

Policy of the Dispossessed King

Sukra also lays down certain rules regarding the disposed king. Once defeated he is to suffer all kinds of humiliation inorder to please the conqueror. He should, then wait for the right opportunity to strike back and regain his glory. He should accept the terms and conditions that may satisfy the conqueror and study the defects and weaknesses of the enemy with the hope of defeating him one day even at the expiry of century. One should wait guardedly like the cat and the fowler and by creating confidence to extirpate the enemy.[73]

Even a dispossessed king should never give up the hope of recapturing his kingdom. A defeated king is advised by Sukra to take asylum in his hill forts in time of great crisis and should make efforts to recapture his kingdom. Sukra allows the defeated king to take resort to violence and even indulge in robbery (dasyuvrtya) to recover his kingdom. It is very strange that robbery was permitted by Sukra. He is advised to be severe, active and unkind in the act of robbery but is advised not to be attached to other wives and violate the girls of respectful families. Later on, during the Maratha rule a class of Pindaris were created for the purpose of harassing the enemy. Sukra says that this robber king is advised not to stay for longer period at one place and should not have blind faith in anybody.

Relation with the Feudatories

Sukra accepts the feudal system where feudatories paid tributes to the king. A detailed account of this system is not found in his nitisara, though some inferences could be drawn. While describing the eight-fold functions of the king, he mentions conversion of princes into tributary chiefs as one of his main functions. This shows the acceptance of feudatory system by Sukra. But, it is not clear what the relation was between the king and the feudatories. It appears that these feudatories were semi-independent rulers who, sometimes, could become so powerful as to force the king to make separate treaties with them in time of crisis. Sukra even mentions of an alliance between the king and his feudatories when attacked by a powerful king. It shows that a feudatory chief could play an important role in dealing with internal and external crisis of the king's state.

Sukra mentions different types of feudatory rulers. The lowest cadre of the feudatories was known as samanta, who ruled over 100 villages with an income of one lac of silver karshas. The next was mandalika who owned 300 to 1000 villages with the respective increase in the annual revenue. The king or raja was one who ruled over 1000 to 2000 villages with an income upto 20 lacs of silver karshas. The maharaja was required to rule upto 5000 villages with an income of 50 lacs of silver karshas at optimum. It, however, appears that the rulers of the cadre of maharaja and sometimes above it are granted the status of independent king by Sukra.

Conclusion

Sukracharya's contribution in the field of military thinking cannot be ignored by anyone. His Nitisastra is a book meant, practically, for all those kings who want to bring prosperity and stability in the kingdom. As such, his work makes him a realist thinker because he realised the importance of power in the political and economic spheres. To keep the subjects secured he even subscribes the use of danda where it is required to apply his intelligence, power, vigour and niti. By the influence of niti, power and wealth can even suppress the sub-human living species.

The foremost duty of the king is to establish good relations with the subjects. In one passage Sukraniti mentions the position of the king. It is quite similar to what Burns has said:

> The rank is but the guinea's stamp
> Through hundred worship at his word,
> He's but a coof for a' that

Shukra wrote: "Does not even the dog look the king a king when it has ascended a royal conveyance? Is not the king justly regarded as dog by the poets?" The idea is that the king is great only from his position but as an individual is no better than other mortals. Hence, the king may command awe and reverence of the people. In case of conflict between the king's officers and subjects the king is advised to take the side of the people. He is to "dismiss the officers who are accused by one hundred men." This truth is not only found in other ancient India literature but also in the present political system. What happens when the king goes against the subjects? In Sukraniti we have the right of direct petition by the subjects advocated several times. Sukracharya's state is a people's state under royal supervision. He is always with the people and for the people.

The next question, then arises, if the king is for the welfare of the people, how are the people to be protected? Here, one finds Sukra's discussions on military administration and art of war gives an insight on the author's formulation of military strategies. His approach could be traditional but it is original. Besides, social and economic welfare, the king always wanted to extend his territories. Even Abul Fazal, the author of Aini-Akbari, devoted certain passages concerning Hindu law. He says: "the monarch should be ambitious to extend his dominion. No enemy is so significant as to be beneath his notice. A wise prince will banish from his court all corrupt men. The king resembles a gardener, who plucks up the thorns and briers and throws them on one side, whereby he beautifies his garden and at the same time raises a fence which preserves his ground from the intrusion of strangers. The king detaches from the noble their too numerous friends and dangerous dependents. In affairs of moment it is not advisable to consult with many."

Similarly, Sukra also discusses military institution and organization in the same fashion. Sukra is perhaps, the only great military thinker to conceive the idea and use of gunpowder in detail, thus denying the allegation made by Alberuni that Hindus were not well aware of new ways and knowledge. His mention of gunpowder has complicated the discussion on the compilation of his work 'Nitisara'. This has been discussed in the beginning of the chapter. Moreover, his preference to other means than war, in resolving the crisis should be appreciated as this is equally applicable in the conduct of modern diplomacy and warfare. His approach to these problems is logical and realistic when he refers to moral and material interests of the king and his subjects but at the same time he also accepted human approach of the king towards his enemy and his subjects. One can readily agree with Sukra's view on diplomacy, interstate relations and war are not outdated in present forms. Like Kautilya, Sukra's military vision influences the modern strategic thinkers.

Though he speaks of king and his method of waging a war but one should not forget the human element in his writings. Today, there are many articles in international law and international institutions which mentions about the war, effects of war, and protection, particularly the humanitarian approach. All these indicate that Sukrachraya was not only well aware of military thinking in ancient India but was equally aware of the changes that could take place in the development of military organisation and institution.

REFERENCES

1. V.S. Apte, *The Student's Sanskrit – English Dictionary*, p. 300.
2. Sukrankti, 1-6.
3. See, U.N. Ghosal, *The Struggle for Empire*, p. 271.
4. Santiparva, 56. 29. 30, 59. 85.
5. P.V. Kane, *History of Dharmasastra*, Vol. 1, Poona, 1946, p. 36.
6. V.P. Varma, *Studies in Hindu Political thought and the Metaphysical Foundations*, Motilal Banarsidas, Delhi, Varanasi, Patna, 1947, p. 107.
7. Santiparva, 67. 16-17.
8. Sukraniti, 1-23.
9. Ibid, 1-4.
10. A.S. Altekar, *State and Government in Ancient India*, Delhi Moti Lal Banarsidas, 1969, p. 29.
11. Sukranitisara 1, 71-72.
12. Ibid.
13. Ibid., ch. 1. Verses 30-34.
14. S.V. Puntambekar, "Some Aspects of Shucra's Political Thought", *The Indian Journal of Political Science*, April–September, 1948, pp. 7-8.
15. H.J. Laski, A Grammar of Politics, p. 21
16. Sukraniti, 1-61 and 4.12.57-58.
17. Beniprasad, *The Theory of the State*, p. 223.
18. Sukraniti, 1.62.
19. B.K. Sarkar, *Sukrniti*, Oriental Books, New Delhi, 1914, p. 2.
20. U.N. Ghosal, op.cit., p. 272.
21. Sukrniti, 1.64-65.
22. Subrniti 2. 214.
23. Sukraniti, 4.2.130.
24. B.K. Sarkar, op.cit., p. 6.
25. Ibid.

26. Sukraniti, 4.1.58.
27. B.K.Sarkar, op.cit., p. 18.
28. Ibid., p. 216.
29. Ibid.
30. Ibid., p. 217.
31. Ibid., p. 217.
32. Sukraniti, 47, 15-16.
33. Sukraniti, 4.7.12.
34. B.P. Sinha, Studies in History of India, Calcutta, p. 118.
35. P.V. Kane, op.cit, p. 679.
36. Sukraniti, 4.7.2-3.
37. Sukraniti, 4.7.16-21-23.
38. Sukraniti, 2.140.
39. Sukraniti, 4.7. 183-184, 174.
40. B.K. Sarkar, op.cit, p. 235.
41. Ibid.
42. Ibid.
43. P.V. Kane, op.cit, Vol. III, pp. 686-87.
44. B.K. Sarkar, op.cit, pp. 235-36.
45. Ibid.
46. A.K. Biswas, Historical Notes, Indian Journal of History of Science, 40.4, 2005, p. 659.
47. Ibid.
48. Ibid., p. 660.
49. A.K. Biswas, Epic of Saltpeter to Gunpowder, Indian Journal of History of Science, 40.4, 2005.
50. Sukraniti, 4.7. 211-217.
51. Sukraniti, 4.7. 235.
52. Sukraniti, 47 284-86.
53. B.K. Sarkar, op.cit, p. 242.
54. Ibid., p. 243.
55. Sukraniti, 47. 356-57.
56. Sukraniti, 47. 358-61.
57. B.K. Sarkar, op.cit, p. 253.
58. Ibid., p. 214.
59. R.C. Majumdar, Age of the Imperial Unity, p. 314.
60. Sukraniti, 4. 1. 19-22.
61. Sukraniti, 4.1.5.
62. Sukraniti, 4.1. 1-2.
63. Sukraniti, 4.1. 36.
64. Sukraniti, 4.1. 23.
65. Sukraniti, 4.1. 26.
66. Sukraniti, 4.7. 238-39.
67. Sukraniti, 4.7. 245-47.
68. R.C. Mazumdar, op.cit., p. 316.
69. Sukraniti, 4.7. 316.
70. Sukraniti, 4.7. 372-373.
71. H.M. Elliot and J. Dowson, *History of India*, Vol. 1, Calcutta, 1953, p. 7.
72. Sukraniti, 4.7. 404-5.
73. Sukraniti, 5.2. - 5.

10

KALHANA

Introduction

History is not only an account of the aspiration of the people for the attainment of political, social, economic and religious goals but it also deals with military development. The political and military developments of a state are related to the rise and fall of empires. Kalhana's famous work, Rajatarangini, is the narration of the kings and royal families of Kashmir. The book covers the rule of various dynasties from 1184 BC to 12 century AD. Though he started writing in 1148-49 AD but he completed it in 1150 AD. It is surprising that Kalhana lays much emphasis on the political and military history of Kashmir where the focuss was on the intrigues, treacherous murders, conspiracies, civil war, tyranny and treasons. It was not something new which he preached. The ancient India is full of these features and as a result we saw the existence of small and great kingdoms trying to overwhelm each other and extending the territories. Kalhana's work is no exception. But he has confined himself to Kashmir.

The credit of introducing Kashmir chronicle to the world goes to Prof Wilson. In 1825, he completed an essay on the 1st–6 cantos of Rajatarangni and published it in Asiatic Society and later on Mr. Troyer took the task of translating all the 8 cantos in 1840 and completed it in 1852. About his work, there are some critics who argue that 1-6 cantos were written by him while the remaining two were the works of others. But Jona Raja, while proving the authencity of the last canto of Rajatarangini, took up the work from Kalhana some three centuries later on and completed it within that span. He explicitly states that Kalhana completed the "Account of Kings" with the reign of Jayasimha. It should be made clear that cantos VI ends abruptly. This this cannot be taken as the conclusion. Hence, it is quite clear that all the eight cantos were written by only one person, i.e. Kalhana.

Kalhana has taken note of both positive and negative achievements of various kings in Rajatarangini. In his treatment, he was influenced by Bilhana's Vikramankdeva caritam and Banabhatt's Harshacharita. Though Bana and Bilhana have focused on the brighter side of the king's achievement, Kalhan, as an objective chronicles seems to have recorded the both sides of the King's performance dispassionately. Objectively, he saw the condition of Kashmir was far from satisfactory. Most of the period with which he has dealt was full of political chaos and confusion. It is also important to mention that most of the kings that ruled Kashmir were assassinated either by the enemies or by some internal defectors. This is well supported by King Jayasimha who said in the occasion on the murder of king Bhiksu: "Alas from Utkaras up to this king, none has departed from this life by natural death".[1]

Before process of analysing Kalhana political and military ideas it is essential to know the topographical importance of Kashmir and then to deduce certain conclusions from Kalhana's political and military writings which made these kings of various periods famous or infamous.

Geography

The strategic boundaries of ancient Kashmir have frequently synchronized with changing political conditions in the state from time to time. Clausewitz has rightly pointed out that terrain is one of five recognised elements of strategy. Terrain included mountain, rivers, woods and road which could influence military activities. The military commander can give proper direction in shaping military force including its size, composition and organization. Thus, geography of a country not only increases the efficiency and morale of the population but also the morale of soldiers whose sole purpose is to defend the territories of the country.

So far Kashmir is concerned, it occupied an important geo-politics and geo-strategic situation since ancient times.

Kalhana gives a vivid account of the topographical feature of ancient Kashmir. The variety of his interest in this field can be very conveniently established even now. Probably, it seems that he had visited each and every place before writing. The exactness of their position and accurate description of Kashmir proves that Kalhana explained geography of the region with precise dexterity.

Ptolemy mentions Kashmir as situated between the Darad Land on the Indus and the Land of the Kulindas on the Beas and further eastward.[2] Hieun Tsang, the Chinese traveller, places Kashmir between Chenab and Indus to the foothills of the Salt range. According to him the valley was surrounded by Urasa in the West, Taxila and Simhapura in the south west and Punch and Rajauri in the South.[3] He mentions the state was 1850 km in circuit.[4] This could have, perhaps, been the extent of the empire of the contemporary kings but not of Kashmir valley which is infact only 500 km in circuit.

Kashmir has a distinct geographical character because it is encircled by high mountains. In the north and the west, the valley is well guarded from the outer world by range after range of high mountains. On the south it is cut off from Punjab by rocky barriers some 80 to 100 km in width.[5] It is interesting to note that how geopolitics influences the relations with frontier states. The application of the term Kashmir valley was generally restricted to the basin of Jhelum and the side valleys drained of by its tributaries about Varahamula (Baramula) defile.

In the south east was situated the principality of Kahsthavata (modern Kishtavar) on the upper Chenab, whose chiefs enjoyed practically independent status. Kalhana's mentions it as a separate hill state in the time of Kalasa.[6] Between Kalasa and Kangra lay the hill state of Champa (modern Chamba). The ancient ruling Rajput family of Champa had maintained matrimonial relations with the Lohara kings of Kashmir.[7]

To the west of Champa and south of Bhadravakasa was the small free state of Vallapura (modern Ballavar).[8] The hilly area including Ballavar called themselves Dogras. Immediately below the Banihal Pass in the territory of Vitasta lived the Khasa (modern Khakhas) with their forts.

For sometime, the Khasa were subservient to Kashmiri rulers but for greater part, they enjoyed subsidies from the rulers of Kashmir.[9]

To the South west and west of Kashmir lay the hill states of Darvabhisara. Actually, it is a combination of Darvas and Abhisaras. This is also mentioned in Mahabharata. The principal principality of this region was Rajapuri (modern Rajouri). Owing to the strategic position of being on the route to plains, the rulers of Kashmir always tried to subjugate it. To the north west for Rajapuri was the territory of Lohara and it included the valley of Lohrin. Lohara seems to have been included in the Kashmir kingdom along with Parnotsa (Punch) in the lower valley of Tohi. Vitasta flowed in between the valley and further to the west lay the kingdom of Urasa (modern Hazara).

Kalhana also mentions the tract of land known as Karnaha (modern Keran). It was under local rule and paid tribute to Kashmiri kings. The valley of Kishenganga was known as Drava derived from Duranda as given by Kalhana. This was a feudatory state of Kashmir and Sarada and one of the most sacred tirthas which is situated therein. It is now under the unauthorized rule of Pakistan. At the other end of the valley was the territory of Dards (Dard-Desa). It was a small kingdom. Thus, Kalhana describes the political geography of Kashmir of his time.

Kalhana felt that it was not the army of Kashmir but the high mountains that have often come to the defence of the valley. Though it true that Kashmir valley is hemmed in between continuous chain of mountains, yet Kalhana has given us a good account of the dvaras or gateways to Kashmir. From these dvaras frequent invasions took place but at the same time the movement of the people was also mentioned to and fro.

However, the modern means of communication both by land and air have nullified the proud claims of the poet. The Pir Pangal range forms the southern and southern western boundary of Kashmir and extends from Banihal pass to the valley of Jhelum. All important trade routes towards Punjab pass through the mountain barrier. The small forts which have been on mountain passes leading into the valley are mentioned in the chronicle as dvaras (gates) or by a more specific term dranga or dhakka.[10]

Kingship

About Kingship Kalhana throws some interesting light on the nature of kingship in Kashmir, i.e. whether it was elective or hereditary. In the Vedic period as mentioned in the Aitereya Brahmana kingship was more or less elective. Kingship became hereditary in the epic age. Ramayana and Mahabharata are noteworthy examples of it. Kalhana's Rajatarangini appears to depict the elective principle among the kings of Kashmir in the initial stages. In the last few verses of the fifth Taranga there is mention of Brahmana assembly who was called to elect the king. There are references of Kamalavardhana, the commander in chief who dethroned the child king Surasarman then the assembly of Brahmanas met to elect a new king. Another instance was electing Yasaskara, a learned Brahmana, as a king of Kashmir. In Rajatarangini it could be inferred that the process of electing a king was not the only work of royal power but it was also dependent on the will of the people in some cases. The process of electing a king was possible when Kalhana mentions that the late kings became tyrant. In some cases, a tyrannical king was killed by the people in an uprising and then a new king was elected.

Kalhana was more concerned with the rise and fall of kings. With the decline of Mauryan empire, innumerable small kingdoms emerged and this made the state an easy prey to foreign invaders. Situated on northern and north-western frontiers, Kashmir, too had to face the foreign assaults. The Mlechchha raided the borders of Kashmir even during the reign of Asoka. These Mlechchha, probably, were Greek raiders who tried to establish their supremacy in north western regions and also Kashmir. Lack of evidential data makes it difficult to reconstruct the history of Kashmir during the period of Greek domination. But one can argue that, most probably, the Greeks continued to enjoy suzerainty over their regions for some 200 years after the downfall of the Mauryan empire till the establishment of Kushana sovereign rule over north western India. After the Kushana, the valley of Kashmir again was ruled by local rulers. Their contribution was in the field of literature and religion. There is mention of dozen local kings who ruled Kashmir. After their downfall, Kalhana mentions the rule of Hun kings. The kings of various dynasties made their presence for not a long time. As a result, Kashmir was without a king for a short time. It was probably ruled by ministers but, later on, Harsha Vikramaditya of Ujjain appointed a Brahmana named Matrigupta as its king.[11] The region of Matrigupta seems to have been of a short duration. On the death of Harsha Vikramaditya, Matrigupta found his position weak. At the same time, he heard that Pravarasena II, son of Toramana, was coming to Kashmir with a loyal band of friends.[12] According to Kalhana, Pravarsena II encountered Matrigupta and being influenced by Parvarsena II, he entrusted him the rein of the government. The kings after him tried to emulate the conditions of Kashmir. Some of them succeeded but the end of Gonanda dynasty came to an end with Baladitya,

Thus, it may be stated that during the remote historical period beginning with the reign of Ashoka the Great, till the rise of the imperial Karkotas, the Kashmir valley was ruled by some glorious kings including a few foreigners. While on the one hand the valley experienced the glorious rule of such kings as Ashoka, Kanishika, Pravarasena and Ranaditya,[13] it also witnessed the barbarous and tyrant rule of Nara Mihirakula etc.

Another feature of Kashmir kingship was that women too in ancient period were competent to occupy the royal throne. Rajatarangini mentions the names of Yasomati, Sugandha, Didda, Jayamati as queen with extraordinary powers as king at the that time and were capable of performing the administrative and military duties.

The late S.P. Pandit[14] has considered in detail the chronology of the various dynasties in Rajatarangini. He mentions these dynasties starting from 1184 BC. The chronological order of the dynasties are given as below:

Group 10.1: Kings of the Gonanda Dynasty

B.C.					Ys.	Ms.	Ds.
1184	1.	Gonanda III	:	Reigned for	35	0	0
1149	2.	Bibhishana I	:	Reigned for	53	6	0
1095	3.	Indrajit	:	Reigned for	35	0	0
1060	4.	Ravana	:	Reigned for	30	6	0
1030	5.	Bibhishana II	:	Reigned for	35	6	0
994	6.	Nara I alias Kinnara	:	Reigned for	39	9	0
955	7.	Siddha	:	Reigned for	60	0	0
895	8.	Utpalaksa	:	Reigned for	30	0	0
864	9.	Hiranyaksa	:	Reigned for	37	7	0

827	10.	Hiranyakula	:	Reigned for	60	0	0
767	11.	Vasukula	:	Reigned for	60	0	0
707	12.	Mihirakula nicknamed Trikotiha, (killer of three crores of people	:	Reigned for	70	0	0
637	13.	Baka	:	Reigned for	63	0	0
574	14.	Kshitinanda	:	Reigned for	30	0	0
544	15.	Vasunanda	:	Reigned for	52	0	0
491	16.	Nara II	:	Reigned for	60	0	0
431	17.	Aksha	:	Reigned for	60	0	0
371	18.	Gopaditya	:	Reigned for	60	0	6
311	19.	Gokarna	:	Reigned for	57	11	0
253	20.	Narendra I alias Khimkhila	:	Reigned for	36	3	10
217	21.	Yudhishthira I	:	No period mentioned			
		Total			**1014**	**9**	**9**

Group 10.2: The Vikramaditya Dynasty

B.C.					Ys.	Ms.	Ds.
169	22.	Pratapaditya I	:	Reigned for	32	0	0
137	23.	Jalaukas	:	Reigned for	32	0	0
105	24.	Tunjina IChange of dynasty ("Anyakulajo Raja")	:	Reigned for	36	0	0
69	25.	Vijaya	:	Reigned for	8	0	0
61	26.	Jaycndra (dynasty ends)	:	Reigned for	37	0	0
24	27.	Sandhimati alias Aryaraja (Jayendra's minister)	:	Reigned for	47	0	0
		Total Vikramidtyas and others, Group 2			**192**	**0**	**0**

Group 10.3: The Gonandas Restored

B.C					Ys.	Ms.	Ds.
24	28.	Meghavahana	:	Reigned for	34	0	0
58	29.	Pravarasena I. alias Tunjina II	:	Reigned for	30	0	0
88	30.	Hiranya and Toramana (dynasty interrupted)	:	Reigned for	30	2	0
118	31.	Matrigupta the poet, (Protégé of the great Vikramaditya of Ujjain defeater of the Sakas) The Gonandas restored	:	Reigned for	4	9	1
123	32.	Pravarasena II	:	Reigned for	60	0	0
183	33.	Yuddhishthira II	:	Reigned for	21	3	0
204	34.	Narendra II alias Lakshana	:	Reigned for	13	0	0
217	35.	Ranaditya alias Tunjina III	:	Reigned for	300	0	0
517	36.	Vikramaditya	:	Reigned for	42	0	0
559	37.	Baladitya	:	Reigned for	37	4	0
		Total Gonandas after the first restoration Group 3			**572**	**6**	**1**

Group 10.4: The Karkota or Naga Dynasty

S.No.					Ys.	Ms.	Ds.
369	38.	Durlabhavardhana alias Prajnaditya	:	Reigned for	36	0	0
632	39.	Durlabhaka alias Pratapaditya II	:	Reigned for	50	0	0
682	40.	Chandrapida	:	Reigned for	8	8	0
691	41.	Tarapida	:	Reigned for	4	0	24
		Total Karkotas up to the end of Tarapida	:	Reigned for	98	8	24
		Grand Total up to the end of Tarapida			**1,878**	**0**	**4**

Group 10.5: The Karkota Dynasty continued

A.D.					Ys.	Ms.	Ds.
695	42.	Muktapida alias Lalitaditya	:	Reigned for	36	7	11
732	43.	Kuvalayapida	:	Reigned for	1	0	15
733	44.	Vajraditya Vappiyaka alias Lalitaditya II	:	Reigned for	7	0	0
740	45.	Prithivyapida	:	Reigned for	4	1	0
744	46.	Sangramapida	:	Reigned for	7	0	0
751	47.	Jayapida	:	Reigned for	31	0	0
	48.	Jajja (brother-in-law and minister of Jayapida's usuprper)	:	Reigned for	3	0	0
785	49.	Lalitapida	:	Reigned for	12	0	0
797	50.	Prithivyapida II, alias Sangramapida II	:	Reigned for	7	0	0
804	51.	Chippatajayapida alias Brihaspati (Son of Lalitapida by a concubine)	:	Reigned for	12	0	0
816	52.	Ajitapida, son of Chippata's brother, deposed and succeeded by	:	Reigned for |			
	53.	Anangapida (son of Sangramapida)	:	Reigned for |	41	0	0
	54.	Utpalapida (son of Ajitapida)	:	Reigned for |			
		Total up to the end of the fourth Taranga			**260**	**5**	**20**

Group 10.6: Change of dynasty

A.D.					Ys.	Ms.	Ds.
857	55.	Avantivarma (son of Sukhavarma, son of Utpala, brother of the concubine above referred to) from (Phalg. kr. I) of 31 to Ashadh. s. 3 of 59	:	Reigned for	27	4	18
884	56.	Sankaravarma, up to Phalg. kr. 7 of 77	:	Reigned for	18	7	19
903	57.	Goplavarama	:	Reigned for	2	0	0
	58.	Sankata	:	Reigned for	0	0	10
905	59.	Sugandha, queen	:	Reigned for	2	0	0

		Dynasty changed					
	60.	Nirjitavarma alias Pangu (grandson of Suravarma). He hardly reigned at all, when he was succeeded but his son, 10 years old, named.	:	Reigned for			
907	61.	Partha, up to Pausa. kr. I of 97, i.e. for 19 yrs, 9 ms., 23 ds., less by 4 ys. O ms. 10 days of Gopala, Sankata and Sugandha	:	Reigned for	15	9	13
923		Nirjitavarma or Pangu again up to Magh. kr I of 98			1	1	0
924	62.	Chakravarma, up to Magh. kr. 1 to 9	:	Reigned for	11	0	0
935	63.	Suravarma, uo to Ashadh. kr. 1 to 11	:	Reigned for	1	0	0
936		Partha again, up to Ashadh. kr. 1 of 11 Chakravarma again up to Jyesh. s. 8 of 13	:	Reigned for	01	511	023
938	64	Unmattavanti, up to Ashadh. kr. I of 15	:	Reigned for	2	0	7
		Total years, Group 5, end of the 5th Taranga			**83**	**4**	**0**

Group 10.7: Change of Dynasty

A.D.					Ys.	Ms.	Ds.
940	65. 66.	Yashaskara, up to Bhad. kr. 3 of 24, including Varnata who reigned a few days before Yashaskara's death	:	Reigned for	9	0	0
949	67.	Sangramadeva, up to Phalg. kr. 10 of 24	:	Reigned for	0	6	8
950	68.	Parvagupta, up to Ashadh. kr. 13 of 26	:	Reigned for	1	4	4
951	69.	Ksemagupta, up to Paush. s. 1 of 34	:	Reigned for	8	6	3
960	70.	Abhimanyu, up to Kart, s. 3 of 48	:	Reigned for	13	10	3
973	71.	Nandigupta, up to Marg. s. 12 of 49			1	1	9
975	72.	Tribhuvana, up to Marg. s. 5 of 51	:	Reigned for	1	11	23
976	73.	Bhimagupta	:	Reigned for	5	0	0
981	74.	Didda, Queen, up to Bhad. s. 8 of 79.			22	9	3
		Total years, Group 6, end of the 6th Taranga			**64**	**0**	**23**

Group 10.8: Change of dynasty

A.D.					Ys.	Ms.	Ds.
1004	75	Sangramaraja, up to Ashadh kr. I of 4	:	Reigned for	24	9	8
1029	76	Hariraja, up to Asahdh, s. 8		Reigned for	0	0	22
1029	77	Ananta, up to Kart. s. 6 pf 39, when he crowned his son Kalasa	:	Reigned for	35	3	28
1064	78	Kalasa, up to Marg. s. 6. of 65	:	Reigned for	26	1	0
1090	79	Utkarsa and Harsa, up to Bhad 5 of 77	:	Reigned for	11	8	29
		Total years, Group 7th end of the 6th Taranga		97	**11**	**27**	

Group 10.9: Change of Dynasty

A.D.					Ys.	Ms.	Ds.
1102	80	Uchchala, up to Paush s, 6 of 87	:	Reigned for	10	4	1
1113	81	Radda a alias Sankha	:	Reigned for	0	0	1
1113	82	Salhana, upto Vais. s. 3 of 88	:	Reigned for	8	0	26
		Dynasty changed					
1113	83 84	Sussala upto Phalg, new moon of 3 including 6 ms. 12 ds of Bhikshachara	:	Reigned for	15	9	27
1129	85	Viyasimha, still reigning in the 25th year or A.D. 1151, i.e. Shake 1072	:	Reigned for	22	0	0
		Total to end of shake 1072, or A.D. 1152, i.e.			**48**	**5**	**25**

Digvijaya King

In Rajtarangi, Kalhana refers to the term digvijaya king. A digvijaya king is one who is mighty and his dominion is stretched beyond his kingdom. A digvijaya king's expansion of his power and prestige depend upon his military capability. Without military power one cannot conceive the consolidation of kingdom. Kalhana has discussed in detail about Lalitaditya Muktapida's achievements in pushing his territories to far-flung areas beyond his kingdom. Lalitaditya Muktapida could be compared to Charlemagne of Europe, Harun-Al-Rashid of West Asia and Hu-Suan of China. Lalitaditya became the important king of the country in the eighth century. He not only held sway over a large area in the northern regions of India but also expanded his rule over a sizeable portion of the subcontinent and areas of the Central Asian regions.

In ancient and medieval periods a king's greatness was mostly estimated by Lalitaditya's conquests. The concept of universal conquest is not something old. Even today a great country is known for the expansion of power all over the world. This was particularly true in the cold war era. In the past, in the west Caesar, Augustus, Charlemagne, Charles V, Louis IV, Napoleon and Kaiser William III believed in the doctrine of universal conquest.

Kalhana's description of Lalitaditya shows that he had a strong urge for more and more expansion of his territories. He is described as a powerful king who had imperial control over India. He brought several trans-Indian regions bordering Kashmir under subjection. He was the most dreaded by his enemies. He spent most of his time in various campaigns. He levied tributes from the eastern kings, probably the Oude and Northern Bihar. He even wore the turban of victory in the Antarvedi or the region between the Ganges and the Jamuna.

After conquering the eastern kings Lalitaditya's first victory was a bloodless one over king of Yashovarman of Kanauj. The king of Kanauj showed himself to be one who knew what was the best thing to do, when he gave his back to Lalitaditya and became his obedient servant. But his allies were more proud than even he was, for the breeze bearing the perfume of the sandal, though only an ally of a flower garden, is taller than the garden itself.[15] It seemed that the peace made between Yashovarman and Lalitaditya was not appreciated by Mitrasarma, the latter minister of peace and war. Even the Lalitaditya's soldiers felt dissatisfied because their desire was fighting a long warfare. But when the allies of Yashovarman showed fight, Lalitaditya took advantage of the adverse advice of Mitrasarma as also of the bellicose attitude of his army. After the unsuccessful efforts of Yashovarman's allies, he deposed the king of Kanauj and rewarded his own minister with five great titles. In this way, the land of Kanauj from the banks of the Yamuna to the banks of the Kalika came under his rule as if it had been a yard attached to his house.[16] Thus, after defeating Yashovarman he established his supremacy in the North.

In Rajatarangani there are certain verses which clearly indicate the defeat of Yashovarman but was spared by Lalitaditya. Lalitaditya, then, moved further and continued on his conquering expedition eastward. Unfortunately, Rajatarangani does not give a detailed information about the further conquests of Lalitaditya in eastern India. After the defeat of Yashovarman, Kalhana mentions his conquests of Kalinga.[17] He marched up to the eastern coast. Evidences prove this in the form of coins of Lalitaditya which have been found at Kanauj, Banda, Faizabad, Varanasi, Nalanda and Monghyr.[18] The big hoard of coins of Lalitaditya found at Banda suggests that the locality probably served as a supply base for his eastern and south eastern campaigns.[19]

So far his move towards eastern India is concered, it appears that his armies did not strictly follow the Ganges's course in the Antarvedi (Doab) but moved in a south easterly direction and crossed the Jamuna.[20] It seems that Banda was of strategic importance as it protected the rear of his army. There was also a possibility of Chandellas making a stronghold in and around the regions. Having consolidated his position and protecting his rear, Lalitaditya then conquered the extensive regions of modern Uttar Pradesh. Coins found at Faizabad and Varanasi indicated that Lalitaditya visited these holy places. After this, he proceeded towards east along the course of Ganges river. According to Kalhana, his progress towards these regions was quite easy because the region extending up to Western Bengal had been included in the empire of Yashovarman.

Kalhana also mentions of Gauda king who was probably the opponent of Lalitaditya. There is mention of Rajabhadra of the Bhadra dynasty who had gained considerable importance in these areas. It appears that Rajabhadra was the most likely foe of Lalitaditya and shared woeful fate in Kashmir.[21]

There exist considerable controversy about the ruler or rulers of Kalinga who had to face the brunt of an attack from the Kashmiri king. After the demise of Harshvardhana, situation in Northern India was quite chaotic. Sailodbhava dynasty asserted their suzerainty in Kalinga. There is great probability that it was either in the time of Madhyamaraja III, the last ruler of the Sailodbhava dynasty or his unknown successor, that the armies of Lalitaditya overran the regions of Kalinga.

Lalitaditya had embarked upon a scheme of extensive conquests and spent many years of his life on expeditions, moving round the earth like the sun.[22] Kalhana gives a detailed account of these conquests which are fully supported by foreign sources.

Being successful in north, Lalitaditya turned his attention towards south up to the river Kaveri. Kalhana's account of southern campaigns are met with certain criticism. M.A. Stein was high sceptical and according to him, the absence of all historical records in detail, as well as the strict geographical order of the countries named, shows that one has in this account merely a conventional elaboration of the popular belief which attributed to Lalitaditya the customary digvijaya of an Indian hero.[23] But there are scholars like Vaidya and Goetz who found no reason to disbelieve Kalhana.[24]

The southern campaign of Lalitaditya has been summed up by R.C. Majumdar : "passing through Karnataka ruled by queen Ratta who paid her homage, Lalitaditya reached the bank of the Kaveri and even conquered some of the islands. Turning west, he overran" the seven Konkans" and advanced as far as Dvaraka in the western extremity of the Kathiawar peninsula.[25]

It may be noted that the campaign of Lalitaditya was not aimed at territorial expansion but was merely a digvijaya befitting a sarvabhauma king.[26] The territories of the defeated kings were not annexed, only tributes were levied and the king was said to have returned to his capital with 11 crores of gold coins from his campaigns.[27]

According to Kalhana, Lalityaditya proceeded further south towards the regions watered by the Kaveri.[28] The armies of the conquering king reached the bank of that river and also the Malaya mountains. These territories were then ruled by the Pallavas. Frequent battles with the Chalukyas of Badami had considerably weakened the Pallavas military force. The Pallavas kingdom was facing internal strikes and it became easy for Lalitaditya to carry his military might upto to the Kaveri and the Malaya mountains.[29]

Lalitaditya also defeated some Silahara king of the South. Here Goetz tried to identify Mummuni of the Rajtarangni with theses Silahara kings.[30] The victory over Mummuni seems to come only after the conquests of the Kombojas and the Tuhkharas in the northern region and before the subjugation of the Bhauttas and the Darads.[31] Kalhana specifically mentions that "....Lalitaditya saw on all sides most of the kings defeated, he entered by pathless tracks the far spreading northern region".[32]

The conditions in the South need further analysis. In the eighth century AD the eastern Chalukyas went through a bad patch because of internal disturbances. As such they were unable to resist any kind of attack. The Ratta queen had the prudence to offer submission. The power of Chalukyas was on decline as Dantidurga, son of the Ratta queen, not only gained more military might but also made friendship pact with Lalitaditya. Seizing the right opportunity, Lalitaditya began to take little by little Chalukyan territories and in the end defeated Kirtivarman II, the Chalukyan emperor. The Konkana chiefs and other rulers were then easily subdued.

Further northwards, there is mention of Maitrakas of Valabhi between the region periods of Siladitya V and Saliditya VII in the eighth century AD. There is greater possibility that Lalitaditya made progress towards these regions. Lalitaditya's army met with little opposition from the Maitraka ruler of Valabhi. Not much is known of these Maitraka kings of Valabhi from 690 AD to 770 AD. Generally, it is assumed that the Arabs attacked Valabhi when Siladitya was in power. These raids took place during 725 to 735 AD.[33] His successor Siladitya VI had to face tough trouble not only from the Arabs but also from the Chalukyas, the Partiharas of Avanti and Rashtrakutas. Here, Lalitaditya does not have to face stiff resistance from them because all of them were trying to assert themselves at the cost of the Maitrakas.

Lalitaditya next target was Ujjayani, the capital of Avanti. Lalitaditya paid a homage to the famous shrine of Mahakala after visiting Dvaraka. To R.S. Tripathi and R.C. Majumdar, the Pratiharas ruled over the regions during the period under review. According to R.C. Majumdar, Nagabhata I ruled there from 730 to 756 AD.[34] There are indications that Nagabhata I did not indulge in any kind of armed clash with a king of Kashmir. Thus, one can easily say that soon departure of Lalitaditya from Ujjain, Nagabhata's I established his supremacy immediately.

According to Kalhana, the route taken by Lalitaditya's army was through the arid and pathless tracts of Rajasthan.[35] Goetz believes that while passing through these areas, Lalitaditya had to overcome small chieftans including the Guhilot ruler, Bappa Rawal who had occupied Chittor (Chitrakuta) and its adjoining areas after defeating the Maurya ruler, Mameraja.[36]

It would, thus, seem that the Deccan and Western campaigns of Lalitaditya were within his reach because of the prevailing political conditions in the South and West. R.C. Majumdar rightly pointed out that these extensive conquests made the kingdom of Kashmir, for the time being, the most powerful empire that India had seen since the days of the Guptas.[37]

But Lalitaditya had to face stiff opposition from Tibet. The seriousness of the threat can be easily recognised from the fact that Lalitaditya had to take the assistance from Chinese emperor. The Tibetans had already occupied the provinces of Yunnan, Szechuan, Kansu and Sinkiang of China. In the initial stage, Lalitaditya had to suffer. The reason could be Lalitaditya's long and arduous campaigns within the Indian subcontinent. With the Chinese king's assistance Lalitaditya was able to check the Tibetan assault and pushed them eastwards. If Kalhan is to be believed then Lalitaditya also occupied the high plateau land, north and north east of Kashmir. Similarly, the Central Asian states' kings met with a similar fate.

The worthy king known for his military prowess and vast conquest met his end on one of his expeditions beyond the frontiers of his empire. According to Kalhana there were two speculations about his death: first it could be excessive snowfall in the region called Aryanaka and secondly he might have committed suicide inorder to be captured by his enemies when he was separated from his main force. But, still his extensive campaigns and victories enabled him the title of digvijaya king.

Council of Ministers

Rajatarangni mentions about the representation of the people of all parts of the kingdom in the cabinet. Thus, the king appointed council of ministers with the purpose of distribution of powers and functions among them. These appointments were generally made because of following qualities viz: administrative ability, bravery and courage, knowledge of the laws and faithfulness to the king. When the king Jayasimha found a revolution could take place in the fortress of Lahara, he soon send Lulla, a councilor of Lahara and Udaya, the warden of the frontier (Dvarapati) in order to crush the rebellion. The king had sent them because they were born in the territory of Lahara and succeeded in suppressing the revolt by watching the local conditions of the scarcity of provisions and other defects.

Loyalty, towards the king was the prime condition for a minister. The most significant example is of Devasaraman, the wise minister of king Jayapida.[38] When he saw his king being made captive by his enemies, he made a secret plan to release him. He took a bold step by showing his inclination that he would pass the throne to the enemy king if he was allowed to meet Jayapida who was well acquainted with the place of secret treasure. He advised the king to jump through the window into the river flowing below where the army was waiting for the king's arrival at the bank. In such a difficult situation, the minister had to lose his life because the king was unable to cross the river and Devasaraman sacrificed his life by strangulating himself to death in order to show that his dead body might help the king in crossing the river. Thus, Devasaraman showed his loyalty by freeing the king from the prison and later on enabled to take revenge against the enemy.

A minister must be wise and advised the king from time to time, particularly in time of danger. The ministers who worked together, should avoid dissension and attitude of mutual jealousy. A slight dissension among the ministers may cause adverse effect on administration. At one place, Kalhana found that such ministers were difficult to find who could boldly tackle a complicated affairs at its close, just as it was difficult to frame the plot of a strong.[39] These remarks indicate that sharp intellect was an essential part of a minister's endowment.

The most important function of a minister was to safeguard his master at critical time. This could be done by giving valuable advice that could befit the time and situations. A good example could be cited from Eight Taranga. Bhoja, descendant of the royal family of Kashmir, sent Nona, his old nurse, to the king Jayasimha with a peace proposal. Dhanya, the minster of Jayasimha, well understood the malacious policy of the enemy. So, he made a plan to test the loyalty of Bhoja. Dhanya met the king Bhoja. There he came to know the intention of Bhoja who was plotting to kill Jayasimha. On his advise, Jayasimha did not accept the peace proposal and in this way saved the life of the king.

So far the relation between the king and his ministers were concerned, Kalhana states that it was cordial. Kalhana had shown how king Avantivarman conducted his relations with his minister Sura. Both were masters in their own spheres when issuing command but were servants of each other while receiving commands. Further, Kalhana states Sura is a real hero who, being aware of the mentality of the king, accomplished necessary beneficent acts without bidding and who does not even spare his life to serve the cause of his lord. Sura represented this rare kind of quality as a minister.[40]

During the reign of Lalitaditya the relationship with the king and ministers were good. Ministers were the true well wishers of the king, respected him much and were free to take decision on the matter pertaining to their departments. The king even could not ignore their well-scrutinized decisions. Minster Mitrasarman, full of self-assurance did not accept the peace proposal of king Yasovarman because in that proposal Lalitaditya's name was placed after Yasovarman. The king, too, supported his minister and declared war against Yasovarman.

With the expansion of territories and increase in responsibilities of the state, Lalitaditya made considerable change in administrative system. The reason, perhaps, could be the accumulation of huge wealth by defeating the opponents and that became a cause of pomp and grandeur of the royal court. Impressed by those who helped the king in performing the function of the state, Lalitaditya rewarded and accommodated them in innumerable engagements.

According to Kalhana, Lalitaditya created five new offices viz. mahapratiharapida, mahasandhivigraha, mahasvasata, mahabhandagara and mahasadhanabhaga. They were called panchamahasabda.[41] The bureaucracy was led by a number of ministers and form a large part of the entire narrative of Kalhana. It has been mentioned that prior to the administrative reforms introduced by Jaluka,[42] the machinery of government was run by following high officials. These were judge, revenue superintendent, the treasurer, the commander of the army, the prohibit and the astrologer.[43] It is believed that though Jaluka introduced the Mauryan system of administration partially in Kashmir, but he did not lay much stress on the constitution of mantriparishad which was well taken up by Kautilya. This was probably due to the prevailing conditions and situation in those regions.

Treasury

Treasury was one of the important pillars of king's functioning. Administration largely depended upon the prosperity of treasury. It was the mainstay of state economy. With proper treasury, it was easy to perform the function of the state and made the people secure and happy. Thus, every king of Kashmir laid much emphasis on the protection and expression of treasury. King Kalasa was quite conscious of it. He says "A King without treasury is liable to be overreached by his own people and by others as well as others. That is why I have safeguarded the royal treasury".[44]

The kingdom in which the people are faithful, devoted and satisfied, the army will grow in strength and it will be united in time of crisis. The wise king who had maintained a proper record of treasury was a cause of great concern for the enemy to seize it.

Forts

Construction of forts was an essential part of the state administration. Forts, whether in ancient and medieval India or Europe, played an important part in defending the territories from the foreign powers. Army stationed in fort could fight the enemy better than those who were deployed in the battlefields. Lothan who wished to secure the throne of Kashmir for himself was well aware of the importance of forts. As the fort in ancient time gave a strong vantage ground in war, it is also became target of attack. King Jaysimha gave instruction to his commanders and officers to capture the fort.

The famous fortress of Lohara is no more but a site covered with stone heaps in the Loharin valley. According to Ferishta, Mahmud of Ghazni invaded Kashmir. He wrote that the fort was remarkable because of its height and strength. Even Alberuni mentions the fort of Lahur was one of the strongest places he has ever seen. Mahmud's army were held up by the seize of Lohara which they failed to take. When the winter approached the season became very cold and the enemy had to receive reinforcements from Kashmir. In the end, Mahmud had to abandon his plan and returned back to Ghazni. As such, Kalhana maintained the importance of fort and observed that those who are the inhabitants of plains can hardly vie with those who live on the mountains.[45]

Military Power

According to Karl von Clausewitz "war is an act of violence intended to compel our opponent to fulfill our will."[46] Violence is seen as a means to achieve political ends, particularly when it is justified on grounds of people's benefits. This has received much strength from Bhagavat- Gita. Kalhana does not lose any opportunity to refer to violence and non-violence on various occasions. But to him non-violence is not a suitable weapon in all circumstances. Hence, violence becomes essential in politics. The reason could be the possibility of revolt by the people against a bad king. Thus, in order to fight out tyranny, it is not wrong to resort to violence. There is no sin in killing such persons even if they happen to be kings.

Violence, was used in case of assassination of Citraratha, a tyrannical king, by a Brahmana youth Vijayaraja. Bhagavata Gita preaches "to protect the righteous, to destroy the evil doers and to reinstate the sacred law, I come into existence in each Yuga ." Violence also act as a deterrent force by the people against the wicked king. For them no royal functionary would venture to oppress people owing to the fear of retaliation from a mettlesome man.

In later times, a potent source of misrule was the rise of the Damaras and the great and small land holding barons. These classes adopted the practice of plundering temple endowments. Kalhana informs that it was with the help of Damaras that Chakravarman won his great victory over the Tontrins. With the emergence of Lohara dynasty in 1003 AD Kalhan's narration shows how the Damaras acquired such military and political influence as to become an unending threat to the royal authority. Harsha made an attempt to eliminate this turbulent class but he had to lose his life. This constant struggle between the central authority and the Damaras, helped by the rise of successive pretenders, continued. In other words, feudal anarchy became the source of conflict. To uproot them, use of force became necessary.

Organisation of Army

In Rajatarangni one witnesses the efficient organization of such a swift and mobile army.[47] Chankuna, a minister of Lalitaditya introduced certain novel features in the organization of army and introduced certain new tactics of warfare which was difficult for the opposing forces to counter.[48] He also believed that the Kashmir forces were well familiar with the Chinese weapons and their methods of warfare.[49]

Kalhana, along with Vaidya, noted infantry and cavalry were the chief elements of army. Elephants forces were relegated to secondary position because of their slow movement. Of the two, infantry and cavalry, Lalitaditya and others paid more attention to cavalry. Chankuna, because of his capability to organize a large army may be given credit in most

Lalitaditya's victory. If we accept Goetz's idea then Chankuna must be given the credit for the introduction of Sasano—Chinese type of cavalry in the Kashmiri army. This proved an added advantage to Lalitaditya in engaging his enemies. The reason for having this type of cavalry was the possibility of threat from Tibet in future and Chankuna was well acquainted with such tactics as to render such attack futile. Just as Ala-ud-din's war machine proved deadly to Mongal hordes and finally turned against Gujarat and the Deccan so the modernised army of Kashmir conquered India until the Tibetans forced Lalitaditya to return to the Himalayan highlands. Thus, Lalitaditya's success depended upon a large and well organised cavalry force. Curiously enough, Bana, writing about a century earlier also was full praise for Kashmiri for horses.

Kalhana's numerous references to the horse and to fox hunting[50] was popular in Kashmir and it also shows that he himself must have been a horsemen. Kalhana has eulogised the skill and interest of Lalitaditya in training horses. Kalhana loves to describe the cavalcade of horseman, the kings who entered Srinagar in triumph and the Samskrta words he uses, Asvavara, still found as Sowar for mounted troops. As in the days of the Crusades steel-clad knights met their antagonists in a melee and in the words of Kalhana "the mighty clash of their arms provided the thunder of applause."

Kalhana's account depicts that the consolidation of Kashmir was largely dependent upon the superiority of his cavalry and large treasure was kept for the purpose of best horses. The best horses came from Central Asian and Afghanistan.

Mobilization of Troops

Kalhana's maintained that the kings of Kashmiri opted both for offensive and defensive types of waging battles. Mobilisation in time, maintaining vigilance at night, tactfully handling the troops during opportune times for occupying positions, withdrawals and the various military plans and not yielding terrain which had been won, such were the merits of the kings which compelled the enemy to flee. In a fashion, Kalhana's praises the offensive action of the king against the enemy.[51] Kalhana also mentions when the army or the king loses confidence because of the treacherous attitude of his subordinates or the king is hard pressed by a furious offensive attack by the enemy, then the king's army should opt for defensive tactics.

Armour: The Greeks have given a detailed account of the Hindu army and its equipments. Each horsemen carried two lances resembling Saunia of Greeks and a buckler. The infantry was armed with a broad sword and long buckler of ox-hide. In addition, each soldier carried javelins or a bow. The coins of Indo-Greek kings and of the Kushans found in Kashmir afford an interesting records of arms, costumes and coiffeur of the period.

In Rajatarangni, one finds the descriptions of chase, tourney, duels, arms, costumes, ornaments and coiffeur. Kalhana mentions that Indian arms were the bow, the long and short swords, the mace, the battle axe, dagger, the katar, chain armour, helmets and visors. Reference is also made of a leather cuirass. It was also popular with the Golden Horde of Chingiz Khan which later on we find in the equipment of the officers of the famous Maratha cavalry. In Rajatarangni, the terms karnirathas has been referred. It is said that Lalitaditya had organised a large and efficient army comprising of 125,000 karnirathas. Though it is difficult to ascertain the exact nature of karnirathas but it is clear that it comprised of a mobile force of skilled archers mounted on karnis and commanded by nobles of proven military acumen.[52]

Military Operation

Rajataragni is divided into unequal eight waves. Each dealing with a single dynasty or a pair of them. Book I consists of 373 verses and deals with the reigns of the lost 52 kings and their immediate successors of the Gonandiya dynasty. Book II has 171 verses and is concerned about some isolated reigns. Book III has 530 verses and there is mention of restoration of power of Gonandiya dynasty. Book IV has 720 verses which covers the Karkota dynasty and Book V has 483 verses and deals with Utpala dynasty. Book VI has 368 verses and mentions the dynasties of Yasaskara and Parvagupta. Book VII has 1732 verses and is concerned with the first Lohara dynasty and Book VIII which has 3449 verses is the longest one and deals with the second Lohara dynasty.

Kalhana's Rajataragni gives a vivid and accurate account of military operations, particularly during the later times. The routes of armies, including the distances and seasons, have been well taken by him. Even Stein supports it. In other words, he gives technical details of the marches, battles and sieges. To make it more effective, Kalhana strongly condemn through the mouth of the illustrious Sahi Trilochanapala (of the Hindu Sahiya dynasty), the rashness and incompetence of his Kashmirian ally Tunga (the minister of king Samgramaraja) in the fight against Hammira, the leader of the Turusha army (ie. Sultan Mahmud of Ghazna).

It was noticed that Tunga gave no thought to night watches, the posting of scouts to military exercises and other preparations for proper attack. The Sahi urged him but of no use to take up his position on the scarp of a hill till he had become familiar with the Turushka warfare. As a result, it led to the defeat of the confederate host. Infact, it brought the descent of the Turushkas on the whole surface of the earth.

In an expedition against the hill-state of Rajapuri during Harsha's time, Kalhana noted how the royal army delayed on the route 'fearing the heat of the Ashadha month.' The success came to royal commander by throwing burning arrows mixed with vegetable oil.

So far the siege of Srinagar by the rebellious Damaras in Susala's reign was concerned it is interesting to note that Kalhana made an indept study how the city was invested by different bodies of rebels from the south (on the bank of the Kshiptika stream), the north (by way of Amaresvara) and the east (on the Mahasarit stream). The roads were kept in uproar with the troops marching out with music, with the return of the wounded, with the flight of routed soldiers and so on. The king showed sufficient courage in defending the capital and personally made proper arrangement for the treatment of the wounded. Not only this, the king encouraged his troops with the gift of marching, giving allowances and medicines and showing gratitude.

Attacking from the east, the rebels were of no match to the king's strong cavalry. They marched by a narrow path along the embankment across the marshes towards north of the Dal Lake. Since they were strong in archers, they showed their best capability in the fight in the narrow embankment across the lake. The king was disheartened by the treachery and mutiny of his troops and disaffection of his subjects. He left the capital for the family stronghold of Lohara by a circuitous route which Stein explains it by a reference to the advanced season.

Equally important was Kalhana's treatment of the new siege of Srinagar by the rebels in 1122 AD, the year following Sussala's restoration to power. More important was Kalhana description of the seize and capture of the rebel stronghold of Sirahsila situated in a most inhospitable region on the north-west frontier by King Jayasimha's forces in 1140 AD. The topographical feature of the area where the fort was situated was important. It was situated between the Sindhu river and the streams of the Madhumati and the Muktasri and the peculiar shape of its hill where it was said to be a "narrow below where it projected into the stream and with a long stretched ridge." Then, the physical and climate conditions of the country along with the areas covered with trees of darkness' and terrible winter due to heavy snowfall made the region difficult to approach.

Stein has identified the site with the Ganes Ghati ridge situated on the Kishanganga which was miles below the shrine of Sharda, now called Sardi. Kalhana and Stein's accounts of the preparations for the siege made by the royal army was fully supported by physical and climatic conditions of the site. The 'Lord of the Gate', Udaya identified by Stein was situated on the same route from the Uttar pargana to the Sarada Tirtha on the Kishanganga. To Stein the region was of great importance because it was the meeting place of several valleys which stretched from the watershed to the Kishanganga. Strategically, it formed an important position for checking the enemy's retreat into Kashmir proper.

Besides this, there was another important general, Dhanya. The beauty of his tactics was building rows of wooden huts for the besieging forces on the bank of the Madhumati. According to Stein, this step was most necessary because the Kishanganga valley had enough level ground only near Sardi. Above it, the land was almost uninhabitable for a considerable distance. Because of heavy rain and snowfall and the extensive forests and numerous snowy peaks the climate was very cold. According to Kalhana, the king sent immense supplies to his commanders because of the inhospitable nature of the region around Sardi. Though the royal forces bravely made presence for three or fourth months, but nothing could be deducted as they neglected to cut off the enemy's food supplies. The forces accepting the direct orders of the resolute king laid a regular siege to the fort.

Leaving his camp on the Madhumati bank, general Dhanya moved to approach the main fort and constructed a continuous lines of block houses. Once the night approached the general kept fires burning so that 'even an ant could not move about without being noticed.' Dhanya further obstructed the enemy's access to water by keeping the boats constantly plying on the river. This helped Dhanya to cut off the enemy's supplies and, thus, prevented them from entering the fort. Ultimately, the rebel Damara leader was so much affected and without food and water he surrendered. Two of the pretenders took asylum to the royalists. The victorious general, then, raised the siege and returned to the capital in triumph.

Military Leadership

In Rajatarangni, the military leadership is analysed as moral and psychological elements of war and indirectly through the medium of genius. The terms genius applies not only to the exceptional individual but also to abilities and feelings on which the behaviour of ordinary man is based. What is important is to analyze those gifts of mind and temperament that constitute military leadership. Taken together they constitute the essence of genius. The military quality as demonstrated by Kalhana was that the king should gain the confidence of the people through an organised machinery of government which could restore normalcy in Kashmir. He should look after the interests of the masses – a task which was neglected by earlier rulers.

Generally, war or battles, are seen as the province of danger. Therefore, courage is the first quality of military undership. Courage could be of two kinds: (i) physical courage and (ii) moral courage. Physical courage is again of two kinds : (i) it may be indifference to danger whether proceeding from the organism of the individual, contempt of death or habits; and (ii) courage may proceed from positive attributes such as personal pride, bravery, loyalty and patriotism.[52] Kalhana has laid much emphasis on these positive attributes.

Bravery

Throughout his work, Kalhana shows high appreciation of heroes death on the battlefield in fighting for protecting the state and its people. Kalhana speaks in terms of high praise of the heroism displayed by Sambhuvardhana and his followers in their unsuccessful fight with the right king Chakravarman.[53] "The valiant Sankaravardhana adorned the bed of heroes together with his dead followers of noble descent and good qualities.[54]

Citing another example Kalhana while describing the complete rout of the allied Kashmirian and Sahi forces against Sultan Mahmud, he praises the illustrious Triclochanapala, 'whom numberless enemies even could not defeat in battle', whose 'superhuman prowess' was remembered even by the victorious Hammira.

Kalhana has also mentioned prince Bhoja, the son of Harsha, who fought bravely against the enemies. He was true to Kshatriyas duties that destroyed his foes like a lion. He felt covered with blood and adorned the couch of a hero and while fighting he felt in combat in a manner worthy of great heroes' envoy.[55] The bravery displayed by king Uchchala before he was overwhelmed by his enemy's act do somewhat atoned says the Chronicles,[56] for the want of pity which he had shown towards the people during his life time.

Of the pretender Bhikshachara, having illustrious career, was betrayed by his followers and was killed while fighting bravely till last. To Kalhana, he obtained in honour the highest place in a great host of kings. He even defied his ever adverse fate at his end by his unbroken vigour and though he was a beggar as compared with former kings, they were nothing compared with him for nobel death.[57]

If Kalhana praises lavishly for his heroes he also condemns the attitude of cowards. He speaks of the cowardly Kashmirian as well as Rajaputra soldiers who abandoned Sussala's body to be mutilated and carried off by the murderous traitors. There are other instances where Kalhana has shown how Kashmirian troops fled from the battlefield in cowardly panic.

Loyalty and Patriotism

Inspite of having a bitter experience and atrocious series of treacheries that stained the history of Kashmir, Kalhana lavishly praises those servants who fought bravely and sacrificed their lives for their masters. He cited the episode of the Gaudas (the people of Bengal) during Lalitaditya period. When the Gauda king was treacherously killed by Lalitaditya , his servants visiting Kashmir under the pretext of pilgrimage boldly attacked the temple of the god who had been made a surety for their master's survival. They destroyed a famous silver image of Vishnu, while being massacred by Kashmirian soldiers.

While taking note of Kashmir, Kalhana's refers[58] in terms of the highest praise to the follower of a Prince who unsuccessfully contended with king Sankavarman for the throne. Inspite these brave soldiers being offered huge wealth and honours brave soldiers were killed in the battlefield for the sake of their own king. Kalhana is very particular to record in his last two books of all those deeds who proved their loyalty to the king amid the general baseness and treachery. Kalhana gives the examples of general Chandraraja, governor Ananda and king Harsha along with Uchchala and Sussala.

With the tragic death of Uchchala's, Kalhana mentions the name of brave Rajaputra, a humble lamp bearer and a faithful Kayastha, who vainly sacrificed their lives to save their king.[59] On the other hand, Kalhana cities another example of a high born official who 'ran away like a dog hiding his danger just as if it were his tail.' On the heinousness treason committed by kings Kalhana characterises traitors as the greatest sinners' and again as ' worse sinners than parricides.' In a latter passage[60] while describing the tragic end of Sussala he condemns one Rajaputra chieftain who disgraced the whole Rajaputra tribe. In the same context, he mentions two brave Brahamanas and a loyal chieftain who alone fell bravely fighting in trying to avenge their master.

Thus, the presence of personal loyalty among the soldiers towards their kings was so strong in Kalhana that he even praises' it even in those who sacrificed their lives in their vain attempt to save the king. To cite few examples, while describing the tragic end of Tunga's murder, Kalhana has great regard to an attendant of the minister who, though unarmed, attempted to protect him with his body.[61] On the other hand, Kalhana had great disregard[63] to other followers of the unfortunate minister who 'sat still in their fright like a women' or else dropping their sword lost both honour and life.

Kalhana also appreciated the value of patriotic sentiment and sacrifices for his own state. He mentioned about the allied Kashmirian and Saha forces who were defeated by Sultan Mahmud of Ghazni. Here, he records[63] the name of three Kashmirian soldiers who fought valiantly in the battlefield and as such 'preserved' the honour of their country from being lost. At another place, Kalhana refers to the capture of Srinagar by the rebellious Damaras during the reign of Sussala. He makes the king conclude his spirited address to a brave companion with noble words:[64] "Is there any person holding a place among self-respecting men who abandon his country at the end without having wetted it with the blood of his body, just as the tiger does not leave the skin without having wetted it with his blood.?"

True to his appreciation of loyalty and patriotism, Kalhana mentions with great admiration of women who stood besides their husband. Kalhana from time to time singles out those Queens who honoured their husband's memory by voluntary death. Even the humble female attendants who took great step was highly praised by Kalhana. He gives an interesting incident of noble wife of the Damara Koshtesvara. When she found husband lying wounded by the king's order she immediately committed suicide by entering the fire. Thus, Kalhana, while referring to the psychological element of war, he particularly made the people proud by instilling these fillings among the soldiers and people so that Kashmirian can be secured internally and externally.

Hunger Strike: A Powerful Method of Coercion:

Hunger strike, to Kalhana, was the most striking weapon in non-violent fight. He mentions as a weapon of the weak against the strong. It was used as a powerful political weapon by individuals or groups, civil as well as military, for redressing against grievance or in defence of the kingdom. It is strange that Kalhana had promoted both types of warfare; offensive and defensive. Later on, in the 18th century Clausewitz made similar statement by referring both offence and defence as a stronger forms of warfare. What Clausewitz meant by offence was a powerful army had to take aggressive action against the enemy and weak army should indulge in defensive warfare.

The western view about the hunger strike is that it is a method of coercion. The Indian views it as not coercive method but to transform the heart of the oppressor or the recalcitrant. The purpose of hunger strike, as pointed out by Kalhana, was to draw attention to some evil or act of gross injustice. In ancient India, the people and sometimes the Brahmanas took resort to hunger strike for political purposes. Even in Rajatarangani we find the prevalence of such an idea. People who were dissatisfied with the judicial decisions thought in terms of unfailing remedy to achieve justice.

According to Kalhana, hunger strike was also seen as a powerful method by the Kashmiri Brahmani to solve the internal problem. In the eighth chapter, one finds them using hunger strike against the corrupt ministers. The Brahmins requested the king to take action against such ministers who were inactive in facing and retaliating the enemies. If no action is taken against them the country would become restless and a prey to enemy's attack. As such, the enemies would take all the autumns crops produced in their land. The Brahmins of Avantipura did not like the rule of Citraratha because of his cunningness which he showed in increasing the imposts. They took resort to hunger strike. Similarly, with the emergence of Damaras in the country, the Brahmins in Vijayeswara were eager to preserve their lands. They began fasting and as a result the king took action when he promised to uproot the Damaras.

Inter-State Relations

It is mentioned in Mahabharta that a wise king should show his valour only when he finds place and time appropriate for it. Force applied at an unsuitable place and time goes in vain.[65] The same comment is also found when minister Haladhara advised the king Anantavarman by saying that "don't recklessly undertake adventurous campaigns in other states."

The Kashmiri king's indulgence in foreign affairs was primarily due to the topographical geography of Kashmir. Kashmir was more or less cut off from the rest of the subcontinent because of its peculiar natural surroundings and mountainous regions. Improvement in the means of communications and great king with brilliance of some of enlightenment policy Kashmir became the centre stage in the management of the whole country. The rise of the Karkotas ushered a new era in the history of Kashmir. The king Lalitaditya not only extended his empire by the use of force in all direction but also maintained diplomatic relations with the king of China inorder to contain the growing power of the Tibetans in the East and the Arabs in the West.[66]

Matrimonial alliances were also used as a tool by the various kings of Kashmir to improve diplomatic relations. Jayapida is said to have married the daughter of Jayanta, a king of Gauda.[67]. This alliance was beneficial to the Gauda king who soon established his suzerainty over the five gauda chiefs.[68] Sankaravarman had married Sugandha daughter of the Darad chief, Svamisena and this matrimonial alliance must have checked Lalliya from coming to the rescue of Alakhana. A similar alliances occurred between the Shahi princess Didda and Kshemagupta. The marriage had far reaching consequences in Kashmir politics.

The famous Utpala king, Sankaravarman seemed to have maintained a diplomatic relations with his neighbours and 'verily made the foreign princes place his order on their jewelled crests'.[69] A more balanced step was taken by establishing friendly relations with neighbouring rulers during the reign of Kalasa. In 1087-88 he convened a meeting of these rulers at his capital to forge a united front. The united front strategy was a novel idea in the history of Kashmir and probably the reason for it was the common threat posed by Muslim rulers. The following kings participated in the assembly were:[70] (i) Kirtiraja of Baddhapura; (ii) Asata of Champa; (iii) Kalasa of Vallapura; (iv) Sangramapala of Rajapuri; (v) Utkarsha of Lohara (vi) Sangata of Urasa; (vii) Gambhirasiha of Kanda and (viii) Uttamaraja of Kashthavata.

It is quite evident from the above list furnished by Kalhana that most of these rulers were halted from the neighbouring tracts. Kirtiraja, most probably, was the same king of Nilapura whose daughter was married to Kalasa. Asata is mentioned in the genealogical table of copper plates of Chamba. Kalhana mentions Kalasa in the context of the expedition of Ananta to Vallapura.[71] Sangramala of Rajapuri owed his royal position to Kalasa when he was threatened by an internecine war.[72] Utkarsha was the second son of Kalasa and held sway at Lohara. Sangata was the most controversial king of Urasa as the various manuscripts give variants of this name. Stein calls him Mungaji which may be taken to mean that his father was Munga.[73] Thus, in Rajatarangni it was seen that the kings of

Kashmir paid little attention towards fostering and maintaining amicable relations with kings of distant lands because of geographical difficulties. Hence, one can infer that their activities were mostly confined in their dealing with their immediate neighbours.

Conclusion

It is a misconception that India did not made any contribution in the realm of political and military thinking. The critics argues that Indian utilized all her intellectual capability in search of metaphysical truth and little military progress achieved in India in ancient times could be deemed as really of sound military importance. But this assessment of India is not true. Military thinking has always dominated the traditions of India and it was well conceived by great thinkers like Manu, Kautilya, Kamandaka, Sukracharya, Banabhatta and others.

Being a highly developed political society in ancient time, the only negative feature was that it was divided into various small and great kingdoms. It should be borne in mind the politics is not the sole criterion of binding these kingdoms together into united organization. More important was the military power which was the means to obtain political objective.

Kalhana's military writing were quite similar to the development that was taking shape in the western world. In both East and West feudal system was the common features. The men of strong breed conquered and ruled peoples and left the economic fight against Nature to others whom in due course they plundered and subdue. Spengler has called this "the breasts of prey and the herbivores." Similar developments took place in India. Kalhana says "there is a natural distinction of grade between man born to command and men born to service, between the leaders and the led of life".

In ancient Kashmir, one saw the decadent of the old economic system and the coming of new system. Kashmir became the meeting ground of different cultures of Asia, the western Greaco-Roman and Iranian and the eastern Mongolian but necessarily it was part of India. As the economic structure collapsed, India became easy prey to internal disturbances and foreign aggressions. Warlords moved and made the life of the people havoc.

Popular rising took place and even Kalhana described Kashmir as a country which delighted an insurrection and they were exploited by military leaders and foreign intruders for their own advantage. In explaining the political and military history of Kashmir, Kalhana, then, gives a vivid account of those chivalrous and mighty kings who with his military power challenged the foreign powers and tried to bring peace and stability in the kingdom. Kalhana also found Kashmir as too strong, due to its mountain fastnesses, to repulse even Mahmud of Ghazni. It was almost 250 years later on, after Kalhana's political and military writings, that Kashmir was subjected to Muslim rule. Even then Kashmir was not conquered by foreign powers but by a local revolution.

Kalhana wrote much before the industrial and technological reached Europe. History to him, was not something to learn but something to make people live and understand life. In his writings he takes all the positive and negative aspects of society and also the virtues of the kings and other characters. In his description of various dynasties, Kalhana emphasises the ability and capability of those great kings who protected the society from internal and external troubles. Great kings were known as digvijaya kings, even though the concept prevailed before him.

Another quality in Kalhana's writings was the freedom from narrow nationalism as we see it today. He pays tribute to the brave man of Bengal who came all the way up to Kashmir and avenged the death of the king who was brutally murdered.

Let us conclude by explaining what place did Kalhana occupies in the realm of political and military thought. It is true and no one can deny that he was a true historian but at the same time he was successful in imparting a high degree of military interest to his promising theme. As compared to earlier ancient Indian thinkers, one can say that Kalhana, no doubt, occupies a pre-eminent position by virtue of his understanding of various military operations. Kalhana's approach of solving political and economic problems through military power was an indication of his extensive study of exact Kashmir topography and chronology of dynasties, his impartial judgment and last but not the least his approach towards the standard of historical criticism.

REFERENCES

1. Rajatarangni, VIII, verse 1785.
2. Ptolemy, VII, I, pp. 21,40-42: vide Stein Rajatarangni, p. 351.
3. S. Beal, Si-Yu-Ki, p. 188.
4. Ibid.
5. W. Lawrence, The Valley of Kashmir, p. 12.
6. Rajatarangni, I, VII, 590.
7. Ibid., I, VII, 1512.
8. Ibid., I,VII, 220, 270, 588.
9. Ibid., II, VIII, 2283.
10. Ibid., 1,1,122;11, 227: V, 39; 11, p. 291.
11. Ibid., I,III, 125-250.
12. Ibid., I, III, 281-320.
13. Ranaditya was also known as Tunjina. According to Kalhana he married the Chola princes, Ranarambha, daughter of king Ratisena. Kalhana is eloquent of his religious catholicity. Due to his nobel character, religious zeal and numerous acts of public welfare, Kalhan compares him with Rama of the Raghu period.
14. The Atharva Veda, the Rig Veda, the Prakrta poem Gaudavaho : compiled dictionary of Prakrta etc, p. 691.
15. Ibid.
16. See K.S. Saxena, *Political History of Kashmir 388 BC–1200 AD*, the Upper India Publishing House, Pvt. Limited, Aminabad, Lucknow, 1974, p. 59.
17. Rajatarangni, I, IV, p. 146-47.
18. Archaeological Survey of India, Annual Report, Eastern Circle, 1919-20, p. 48.
19. R.S.Tripathi, History of Kanauj, 1937, pp. 203-04.
20. Rajatarangni, I, IV, p. 132.
21. Journal of U.P. Historical Society, XIV (NS), Pts-1-11, pp. 63-70.
22. Rajatarangni, I, IV, p. 131.
23. K.S. Saxena, op.cit, p. 64.
24. See C.V. Vaidya, History of Hindu Medieval India, 1, p. 207ff, 300 ff, H. Goetz, Journal of Bombay Branch of Royal Asiatic Society, Vol XXVII, pp. 43-60.
25. R.C. Majumdar, the Classical Age, Bharatiya Vidya Bhawan, Bombay, 1954, p. 134.
26. Rajatarangni, I, IV, pp. 126, 183, 189.
27. Ibid, pp, 176, 189.
28. Ibid, I, IV, p, 155.
29. Ibid, pp. 155-56.
30. Goetz, op.cit, p. 49.
31. Rajatarangni, op.cit, pp. 165-69.
32. Ibid, p. 163.
33. R.C. Majumdar, op.cit, p. 150.
34. R.C. Maumdar, Age of Imperial Kanauj, Bhartiya Vidya Bhawan, Bombay, p. 20.
35. Rajatarangni, I, IV, p. 163
36. H. Goetz, *Journal of Bombay Branch of Royal Asiatic Society*, XXVII, p. 57.
37. R.C. Majumdar, *Ancient India*, Motilal Banarsidass, Delhi, Varanasi, Patna, 1964, p. 262.
38. Rajatarangni, VII. 449.
39. Ibid, VII, 1012.
40. Ibid, V, p. 62.

41. Ibid, I, IV, pp. 140-43.
42. Ibid, I, I, p. 118.
43. Ibid, I, I, pp. 118-119.
44. Ibid, VII, p. 990.
45. Ibid, VIII, 2255.
46. K.von. Clausewitz, On War, translated by J.J. Graham, London, 1949, p. 1.
47. Rajatarangni, I, IV, 246.
48. H. Goetz, op.cit, XXVII, pp. 47-48.
49. Ibid.
50. Rajatarangni, VI, 183.
51. R.S. Pandit, Rajatarangni, Sahitya Akademi, Delhi, 1990, p. 583.
52. K.S. Saxena, op.cit, p. 80.
53. Rajatarangni, V, 336.
54. Ibid, 337.
55. Ibid, VII, 1655-56: 16-76.
56. Ibid, VIII, 331.
57. Ibid, V, 1768-71.
58. Ibid, V, 131.
59. Ibid, VIII, 323-26.
60. Ibid, VIII, 1325-27.
61. Ibid, VII, 82.
62. Ibid, 87 FF.
63. Ibid VII, 59.
64. Ibid VIII, 1192.
65. Mahabharta Santi, 140/28.
66. U.N. Ghoshal, *Studies in Indian History and Culture*, Calcutta, 1966, p. 177.
67. Rajatarangni, I, IV, 461-62.
68. Ibid, I, IV, 467.
69. Ibid, I, V, 138.
70. Ibid, I, VII, 587-590.
71. Ibid, I, VII, 220.
72. Ibid, 533-35.
73. Ibid, I, VII, 588-90.

INDEX